The Austinl Brothers

The adventures of two Army Brats

By

David Austin

Table of contents

Chapter 1
Born to the Military:
A Life Without Borders

In the years of our Lord 1954 and 1955, two boys were born into an army couple. The following is the life, times, adventures, and trails of those two brothers or, otherwise known as two military brats. The definition of a military brat is a child who has a parent in the military. It is somewhat of a subculture and a lifestyle identity. The lifestyle involves typically moving to different states and countries often while growing up. It is customarily for a military person to be transferred to non-combat assignments to where the military person can take his family along with him. Most of the time, a military brat never has a hometown due to the constant moving around. War-related family stresses are also a commonly occurring part of the military brat's life due to their parent being transferred to a combat area. Many feel that military brat's mobility experiences in their upbringing give them a sense of worldliness. Most military brats like the term but outside the military world, the phrase can sometimes be misunderstood by non-military folks as insulting and abusive. To some, a military brat is a cast out in the world and the non-military doesn't want anything to do with them. But to us military brats, it gives us great pride and confidence in handling life's many problems. As my brother and I would say, "We have seen the world and would not change anything for it."

Many will think that what has happened to these two brothers is impossible and that it could have never happened. We don't really know why we did what we did or why but we both lived our lives to the fullest. Most of our advantages could cause body injuries or death, but we both survived and have the scares to prove it. Those reading this book might think I am writing in a strange language. We were raised in a different decade and with a southern flare, so you may not understand what is written. I have tried to keep the writing as if the words were spoken because that is the way we live. I guess the only way you could fully understand how we lived our lives during that time frame was that you had to live it, and my little brother and I lived it to the fullest extent of our lives.

We were raised in the South for a portion of our lives and learned that Southerners are hardworking folks who love their trucks. It may be somewhat of a stereotype, but we do love our trucks. Southerners hold on to their identity with a vexing stubbornness—but really, we just love who we are. The reason we hang on to this style of life is that we love those who came before us and raised us to be like them. We learned at an early age that by saying y'all and bless your heart, to say if I had my druthers, hold your horses, saying yes ma'am and yes sir, you got a lot further in life. By respecting our elders and treating them with respect and dignity that it just makes for a better life. Southerner males are taught at a very young age that you should always open doors for women and girls. NO, it is not old fashioned but is being respectful. You also were taught to call folks Mr. or Mrs. And to us their last name, not their first, again a show of respect. Being Southerners, we were instructed on how to hunt and fish and how to take care of ourselves in the woods or on the water. We learned that today would frown on like drinking water from a water hose, riding in the back of a truck, taking a bath in the lake, and eating berries right off the bushes without washing them and eating grits, not cream of wheat. To pick garlic grass from your back yard and us it for cooking. We learned at a young age how to clean a buck without spoiling the meat, how to run a trotline without embedding the hooks in your hand, how to dress a squirrel, how to pop the breast out of a dove and how to skin a catfish by nailing it to a post and pulling the skin off with a pliers.

We learned a lot of old remedies and have tried to pass them on so that they are not forgotten. My grandma showed us so many of these remedies, from how to treat a bee sting with wet tobacco to how to put a piece of salt pork on a boil to pull out the poison.

Mom taught us how to cook our spoils and we ate what we killed. We did not believe in killing for sport but to keep food on the table. One of the big things that we will do is Mom would save the grease from the bacon we cooked and use it to fry our game. Bacon grease just adds something when cooking, and we all know that you can't go wrong with bacon! You were in hog heaven when she would cook fried squirrel with mash potatoes and homemade brown gravy with yeast roll. The only thing you had to watch out for was the buck shot from the shotgun that would be imbedded in the squirrel. If you have ever bitten into a fried squirrel and hit a buckshot, you know that felling! Felt like you just cracked a tooth and sometimes this was the case. We also had to have some kind of sweets in the house, from homemade cookies to pies or cakes. My Dad had to have something sweet to eat after lunch or dinner, normally after dinner. So, we had cakes and pies and all kinds of cookies made from years of hard work and experimenting by our grandmas, great-grandmas, and from the ladies in the church.

Our grandpa was from an old generation where you didn't buy the stuff you made it. He would roll his own cigarettes and use old filters in them. He would buy a pack of smokes and then save the filters after smoking them. He would keep using them over and over when he rolled the new ones. He also drank Dr. Pepper like it was water. Every time we came over, he would have a Dr. Pepper in his hand. For every meal, he would drink Dr. Pepper, too. My brother and I learned a lot from our grandpa. He could fix anything and you never touched his tools. We got into his work shop one day and started playing with his tools, and he caught us. He whipped the hell out of us and just kept saying, "You do not touch my tools!" One of the things that my brother and I can still remember is he taught us how to make a sidewalk. Showed us how to make the mix to make it strong using sand and a concrete mix. He taught us how to put the frame together, pour the mix into the frame and clear the top. I will never forget the experience.

Both of us were raised in a church environment. You went to church no matter what unless you were at death's door! Church in the south can be a complex ordeal. Just choosing a church has some back ground in the south. There are two types of churches in the south: contemporary and traditional. Worshippers of the contemporary type are fine with wearing jeans to church. A contemporary church will use instruments like drums and to keep your hands free so that you can clap during the songs and sermon. Traditional churches want you to hold your hymnals in your hands so that there is no clapping cause clapping makes folks nervous. These folks felt that a piano and or an organ are all you need to be biblical and you always wear your "Sunday best" cloths to church. Ours was a backslapping, loud singing, and a true family experience. Our church has always had your back. If you needed your lawn mower or windows cleaned, if you need grousers and a ride, they were there for you. Also, pew selection is a thing in a traditional church; you have YOUR own pew, and you always sit in that pew! The reason I mention this is that in Arkansas, when dad was on leave, we went to The First Baptist Church of Hoxie. If you were to go outside and look at the corner stone of that church, you would see my grandpa's name, and he was one of the original builders of the church. To this day, my mom sits in the pew that our family has been sitting in for decades. This is where many lessons were learned and the rules were set down by my dad. It all stemmed from respecting your elders and helping each other out in bad times. My brother and I learned a lot and carried those understandings and beliefs over into our adult lives with our families.

As we got older, we were still astonished that we were still "alive and kicking." Everything in this book happened and we both are proof that it happened again, we still have the scares to prove it. There is no way you can make this stuff up, but it really occurred.

Also, we are dedicating this book to our father, our old man. There is a country song out there titled "My Old Man," and if you listen to it, you will see that it is the way our father was. He was a hard worker, and he played just as hard. He disciplined us the best way he knew how mostly with a belt or anything he could fine to bust our butts. When you have kids, you are not given a book that tells you how to raise your kids or how to keep them from harm. You do the best you can and when times require discipline, you discipline your kids. The world we were raised in was you did not "spare the rod" but wiped butts when needed. One day, my mom, out of the blue said that your dad would have been put in jail for what he did to us boys and to that, I replied that we lived in a different time and that was the way the world revolved.

My Dad made the Army his career. He was in the Army for 27 years, retiring as a Command Sergeant Major. Throughout those decades, he saw the world change right before his eyes. He was deployed to the Korean War near its end, and he took a tour of Vietnam.

While in the Vietnam War, he was also involved in the Battle of Hue. The Battel of Hue changed our dad and most of the time, he would take it out on us, his boys, mostly to make sure that we did not do those things that he saw during the war. He developed a drinking problem and a smoking problem but he overcame both of them. He went cold turkey with his smoking and one day, he resolved that he would give up the alcohol and dedicate his life to Jesus Christ!

The time frame of this book also drives the circumstances of what happened to us. During this time in this span of our lives, discipline was very severe and firm, so there was a lot of "ass whippings" that took place. Most were administered through love and discipline; others were administered through the effects of alcohol. Of course, we deserved most of the "ass whippings" we got but through it all, it made us tough and got us through life. Being an Army brat is not easy. Life in the Army or any branch of the military is completely different from that in the civilian world. You move so much and you have to make friends as you go. You hope you get to stay in a place long enough to make friends. When we grew up, my brother and I enlisted in the military. We Both enlisted in the US Air Force. My brother was in the Air Force for four years;

I, on the other hand, made a career out of it, as my dad did, and was in for over 20 years.

With us averaging a move every 24 to 30 months per assignment, there really wasn't time to make friends. Our Dad was stationed within the United States and in Europe, so our travels were worldwide. So, my brother and I became the "bestest of buddies!" We would take on the world together and most times pay for it in the end, no pun intended. We did more things together in my dad's twenty-something career than most folks would do in an entire life time. Mom told us that Dad said wherever we got stationed that he would take his family on an unusually exciting activity to explore the local area and discover the area and what it provided. Where ever we were, within the United States or in the European Theater, we would travel and take a look at the country. Most of the time, there was nothing that was off-limits to us boys. Those that were off limits, according to our fathers, would be paid for by a strap across our butt or a slap across the face. Oh yes, we deserved most of the beatings we got but our lives were so full of adventures throughout the globe. And through those ass whippings, much deserved in many instances, it made my brother and I into who we are.

We are both in our late 60s and early 70s now and keep asking each other if we knew what we knew, would we change our lives? Would we have taken better care of ourselves? We both agree without hesitation that all through all these adventures, we would not be the tough, rugged, resilient, loving brothers we are today. As far as the damage we did to our bodies during those years, well we are paying for that every moment of our lives now but still living our lives to the fullest, as best as our bodies will allow.

Chapter 2
Arkansas

When our father was deployed to Korea, we were in the First and Second grades in the early '60s. We were at our home in Arkansas while he was deployed to a combat zone. This home was bought by our parents when Dad first enlisted in the military. One of our favorite things to do at home was when it got cold, and the creeks and farm lands froze over, we would get on our bikes and go riding on the ice. Our game was to see how many burses we could get. So, we would ride the bike on the ice, and when it got to a thin patch, most times, the front wheel would crash through the ice and stick, which would flip you over, and you would go sliding across the frozen water. We would stay outside till dark and just ride our bikes. During this time frame, we lived on our bikes. We road them everywhere. One good thing was that we never broke a bone doing this but we had so many burses that mom would worry about us.

We also had two big oak trees in the backyard. This is where we build our tree fort. The trees back then were about 30 feet tall and about three to four feet around. We got some old wood and built the steps up the side of the tree to the big fork about 15 feet from the ground. We would lay the boards over the limps and we even put a tent up there. During this time frame, Superman was very popular. That would be the black and white series of Superman. We would sit and watch it on our blonde colored black and white TV. One day, we got an idea: why don't we get a sheet, tie it around our neck, hold on to the corners, jump out of our tree fort, and fly to the ground? This was not a great idea. Of course, me being the oldest I had to go first. I jumped out of the tree and went straight down, landing on my side. Oh man, that hurt, but it did not break anything. So, we thought that with all the leaves the oak tree was producing, we would pile them up and jump into the leaves. So, my little brother took the next jump and again this was not a good idea. He hit the leaves and they flew everywhere. He landed on his side too and man he was hurting with only burses. Our pump house which was about 8 feet off the ground, was about 10 feet from the tree and many times we would take a little run from the tree fort and jump, hitting the side of the pump house head-on. That too, was a very painful experience. After a while we figured out that we could jump from the tree and land feet first on the pump house and then jump again to the ground.

Another adventure at our house was to go out to the woods, ride the small trees to the ground, spring back up, and fly in the air. Our house sat right outside the city limits and we have a very large wooded area that surround our house on three sides. Our house was surrounded by cotton fields or soybean fields, depending on what was in season. We would go out to the trees and jump on the small trees which were about five to six feet tall and about three to four inches around, and ride them down to the ground. The tree would spring back up, and we would just fly like Superman. Most times, the tree would spring back up, but sometimes they would snap like a toothpick, and you would end up face first in the dirt. Mom did not care too much for us breaking the young trees. She would come running out of the house and tell us to stop riding the trees. Well, that is just like saying, do it so we just keep on riding. She came back out after a few times shouting at us, very pissed off, and told us she was madder than a hornet's nest and that we needed our butt's whipped for not listening to her. She told us to get her a switch so she could whip our ass. My little brother being the smart ass that he was, got here a whole tree about three feet long and gave it to her. Of course, this only infuriated her more! There were some willow trees in the back and she went over to the tree, broke off a branch, and began to whip us all the way up to the house. It must have been the length of a football field from the woods to the house and she did not miss a lick. A willow branch can reach all the way around your body and she just keeps whipping us. We had some pretty good welts, had stripes like a zebra all over our backs and butt. Needless to say, we did not ride the trees anymore.

The wooded area in the back of the house belonged to one of the local farmers. We would go out into the woods and play all the time. One day, my cousin, my brother and I went out to play and we were pretty deep in the woods when we noticed a small wire running around the trees in a large area of the woods. The wire was about three feet above the ground, and for the life of it, we could not understand why it was there. We were out there just a few days before, but it wasn't there. We followed the wire around a whole cluster of trees and noticed it was attached to a car battery which was covered in a small tin box over in one of the corners of the big squared-off area. Being young boys and very curious about the wire, we grabbed hold of it. Wholly cow, that thing shocked the shit out of us! After we got up, we all laughed our butts off. We got to playing army and running around when we all three had to take a piss. Now, as young boys, you pee together and of course, being as completive as we were, tried to pee as long as we could to be the last one to finish. We were peeing pretty good and moving around, trying to pee on each other, and we all hit this one tree. The only problem was that this tree, out of all the hundreds that were around us, had an electric wire running around it. OH yes, our pee hit it, and we got the shit shocked out of our penises! OMG, that hurt, and of course, as you are peeing, you can't just stop so it lasted a good ten seconds peeing on the wire. We were rubbing our privates for hours trying to get feeling back into our penises!

One more time, while in Arkansas, I got into a fight with an entire swarm of big black bees. I was about 12 years old and my buddy and I were out throwing rocks at bottles and cans when we came across an old car that had been turned over on its side. The windows were just calling for us to throw rocks at them. We started pelting the windows with rocks. We broke every one of them in no time. Now, we were curious about what was in the car so we walked over to the car. I was the tallest of the two so I climbed up and stuck my head in the driver's door. Oh lordly, I had just stuck my head into a hive of these big black bees, the ones with the black bodies with yellow stripes. They must have been two inches long and looked like a whole army of them. I jumped down and started running for the woods. I did not make it. I figure about five or six of them got down my tee shirt, not sure how that happened but they began to sting me. When it was all said and done I had more than five big welts, stings, on my back. They swelled up to the size of your fist and it hurt like hell. I ran over to my grandma's house, which was about two blocks away. I was crying my eyes out and told my grandma what had happened. She ripped my tee shirt off and saw all those welts. She got some tweezers and started pulling out the stingers.

She counted 5 stings that were embedded in my back. She then got some of my grandpa's smoking tobacco, soaked it in water and started to place it over the bee stings. I didn't know it but tobacco juice acts as a substance that will pull out the poison from a bee sting. She put some band-aids over the tobacco and told me to drink lots of water. My back hurt for a few days. I guess this is one way to figure out that you are not allergic to bee stings!

At our small house in Arkansas, we had to pump water out of our well which was located in the backyard. It was a hand-pump water system that was near the house. One of the major rules that Dad sat down on was that you would always keep a cup of water near the pump after every use so that you could prime the water pump for the next time. My brother and I both learned the lesson of the water cup for the pump the hard way. Young boys forget a lot, and this was one of those occasions. Numerous times, we would go out and get water for the chickens and forget to keep a cup of water for the pump. And it never failed that Dad was the first one to go pump water without a cup of water to prime the pump. This really pissed him off, especially after about the fifth or so time that he told us. You would think that after the first ass whipping you got for not filling the cup, you would learn your lesson but no, not us. I bet it took more than five ass whippings but we finally got the point and started leaving the cup of water. I guess I need to let you know too that if one boy got an ass whipping, the other one did too, no matter what. Dad would say that it was a learning experience and that both boys needed to learn. It didn't matter if my brother did something or I did something. We both got our ass whipped!

This might be the time to mention that our dad loved to barter. He was very good at trading a service or goods in exchange for something that we needed or he wanted. Most of the time, the deals did not involve money but services. My Dad was a professional butcher, and he worked many years in the meat department at our local grocery store in town when he was growing up. Most of his deals transpired when it was deer or elk season. It was not uncommon for Dad to wake up and find a deer carcass hanging in the barn, only to get a phone call asking if he could cut up the deer. The majority of the time, his standard rate was that he would take one-third of the meat. Dad would cut, grind, and package it for his customers.

Sometimes, if the time of the season was right, he would get a bounce from his customers in the form of an ice chest full of pears or apples, which he and mom would make into preserves. I remember when our fruit trees were coming in in Arkansas, and one of his old buddies came over and made a deal for the fruits. One time one of his buddies brought over an ice chest full of crappy fillets. Now, this is one of mom's favorite fish to eat; she just loves it. She would tell you she should eat her weight in crappy! With all these deals our freezer would always be filled with food, mostly from dad's bartering.

We seemed to always have a truck in our family. Dad had acquired an old 1959 Ford five window pickup that ran sweet. He found it on the side of the road in someone's front yard with a for sale sign on it. It cost him $500. This Ford had a push button start on the left side of the floorboard but it didn't always work so many times you would see us boys push starting the truck while Dad sat in the driver seat. That truck ran like a top. Just put gas in it and ensure the levels are good; she will last forever. Dad would use that old truck for everything.

We can remember one deal he made with the local sawmill. We had a wood-burning fireplace that Dad had created and had it built in the den. And he would always have a fire going. He did not like splitting wood so he hooked up with the owner of the sawmill. They would take a large tree and cut the ends off, which they called slaps. They would stack the slaps up and, at the end of the day, would burn them. So, Dad approached the sawmill and asked about the slaps that would pile up. The owner stated that they were not good and that they burned them every day. Dad said what if I came over every so often and got a truckload of those slaps? Would that be OK? The owner said, "Knock yourself out." I guess it isn't called child labor if it is your own sons doing all the work but my brother and I would load that truck so full of wood slaps that the front wheels would come off the ground. Now that is a load! The only thing is trying to drive that old truck without the front wheels on the ground. It would bounce up in the air and come down, and at that time, you would make a turn. You had to judge it so that once it hits the ground, you make a small change to keep it on the dirt road. Oh, it took us hours to get home, but Dad got his wood. Of course, Dad would say that you boys loaded the truck, and now you get to unload it!

We had that truck for years until Dad was tired of the upkeep and wanted to sell it. He would tell us boys that he bought it for $500, so he was going to sell it for $500. He had owned it for over 5 years but still wanted what he paid for it. He parked it in the front yard with a sign on it and many folks came by and asked about it. They all wanted to pay less, always asking what he wanted for it. It never failed; he replied that the sign says $500, which is what I want for it. They would say that they would give him $300 or $400 for it and he again would say that I guess you aren't buying my truck because it is $500. He stuck to his guns and finally sold the old truck!

So, this left us without a truck. So, the search began for a new old truck. Dad found one later that year when he sold the Ford. It was a 1985 Chey pickup. We named it Big 10. At the time of this writing, we still have Big 10. It had a new engine put in it a few years ago, and it hasn't been running in over five or six years sitting in our sister's backyard. My brother has made arrangements to get it transported to his place in Florida where he is going to get some buddies and make it all original, and pass it down to his son.

There was another thing about our Dad that really stuck with my brother and me and still does to this day. If you ask someone, they will tell you that they have a favorite place to sit, either on the couch or in a chair, and that is their favorite spot. Well, Dad had his favorite; it was an old recliner. He loved it and slept in it numerous times, but one big rule was that you never sit in his chair. If you forgot or were new to the house and sat in it, you quickly found out it was his chair. His favorite saying if you sat in the chair and he was in the room was, "Let's play checkers, and it is your move. Get the hell out of my chair". This may not sound like much to the reader but to this day, even after Dad has passed on my brother and I cannot sit in that chair. In the living room we had to move the chair to a different spot and only mom will sit in it to this day!

One adventure we had was taking Big 10 to the rice fields surrounding our house. The fields would be flooded so the rice could grow, which would bring out the big bullfrogs. We had a good system for getting the bullfrogs too. One of Dad's old friends, a retired Colonel in the Army who could make anything, designed a 10ft frog gig. He made the pole out of oak about two inches around, and the gig was extra-large, with five prongs that were about seven to eight inches wide. Dad's friend would drive the truck down the dirt roads between the rice fields once they were flooded. I was in the back of the truck in the bed with a headband flood light. Dad would be in the passenger seat with a handheld light. I was in the back of the truck with the monster gig and a light on my head. Dad would spot the frogs with the floodlight, which I could see from the back of the truck because the light would reflect off of their belly. Everyone knows that when you put a spotlight on a big bullfrog, he freezes, enabling you to gig them. Dad would spot the frog, and his friend would get as close as possible on the side of the dirt road without going into the rice field, as close to the frog as possible, so I could gig it. I would then put my light on the frog and grab it from the back of the truck. The back window of the truck had a sliding glass too. Why am I saying that; well, that is how I passed the beer from the cool in the back of the truck to Dad and his friend. The reason for this story is that there is always a learning experience in just about everything we do in our lives! Dad always seems to teach us boys the easy and hard ways of life, which was one of those times. I gigged it when we came up on the first frog but good. I think I got all the prongs in its back; that is how big the frog was. As I was bringing it into the back of the truck, Dad said that I needed to step on it to hold it down or it would jump out of the truck. Well, I was a young, immature juvenile and I knew everything and there was no way that a bullfrog with all those prongs stuck in its back would even be alive, much less jump out of the bed of the truck. As I brought the frog into the truck, I bent over and pulled the gig out of it. As I was bending over to pick it up once the gig was out and put it in the burlap sack we used to carry the frogs in, it jumped clean out of the truck, a good 10 feet. Dad said some words I can't put in this book, repeated his statement, and asked for another beer. From then on, I put my foot on the frog as I bent over to pick it up so as not to lose them. Once we got home, we had to clean them, but me being me, I took a big frog into the bedroom where my wife was sleeping and woke her up and shoved the frog into her face, to where you could hear her scream a few counties over and got slapped so hard I dropped the frog. Now I had to catch him. He jumped all over the bedroom but I finally corralled him. This led to another teaching experience from Dad on how to clean a bullfrog. It is said that if you do not remove the cord that leads down the length of the leg, the leg will jump out of the pan once you put it in hot oil.

Again, me being the young juvenile, I did not believe Dad. You guessed it, I put the very big frog leg into the hot oil and that son of a gun jumped right out of the pan on my chest. Did I say it was hot oil? I got burned, but I lived and my dad just said I told you so. So, from then on, I removed the cord!

On another occasion, when Dad was deployed to Vietnam, we went home again to Arkansas. At this time in our lives, we were going to school in our little town and we were in the fifth and sixth grade. Both of us can remember that the commercial fishermen who worked Black River would bring in catfish to show off to the kids in the school. We can remember them bringing in catfish that they had caught that day. Caring them by the gill slits, which is the opening between the gill's arches of a fish with both their hands in the gills, lifting them up over their head, and the tails would still be dragging on the ground. The fishermen walked into our classroom, bragging about their catch. Those had to be the hugest fish we had ever seen.

We would often stay at my grandmother's house, which was about half a mile from our house. My grandparents ran a small hamburger place that they called the Pink Castle. It was located just off the school grounds, and all the kids would come in and get a burger, chips and most times, a bottle of Dr. Pepper during their lunch break. During that time, my grandfather developed a way to pick up bottles that the kids would throw away on their way back to school. He would take a paint roller handle, slip it over a broom handle, nailed it to the end of the handle and pick up bottles with it. You would slide the end into the hole of the bottle, pick it up, and put it in the old wooden crate.

My brother and I would get our red wagon, put the crate in the back of the wagon, along with our handles, and we went collecting bottles from the ditches and sidewalks after school. My grandfather would pay us 5 cents a bottle. We could collect a whole crate most days. It not only gave us some money to spend, but it also helped to save my grandfather's back.

One day my brother and I strolled down to our town's water tower. Every small town in the south had a big water tower that normally had the city name on it, and nearly all of them were located in the middle of town. Our tower was about as high as a five to six-story building. There were steps leading up the tower and at the bottom of the water container, there was a walkway with a railing that reached all the way around the container. We would climb the tower, sit up there most of the time with our lunches, and eat during lunchtime at school. We would drop stuff off the tower and see how far the wind would take it. We also would stand up by the rail and spit off the towner. We wanted to see if we could hit anyone on the ground! There was this one particular day we had climbed up the tower. Mom had told us numerous times not to climb the tower but that is like telling two young boys to climb the tower. On this day, Mom was helping our grandma feed the school kids during the lunch rush at the Pink Castel. The tower was only about six blocks from the Pink Castle, and wouldn't you know it Mom saw us up at the top just sitting there eating our lunch. She came storming down the road, screaming at the top of her lungs to get down. Didn't I tell you not to climb up there? She then followed all that screaming by saying that she was going to whip our butts. We were both scared to death because we had never seen Mom so mad and climbed down the tower. Once we got to the bottom of the tower, she was standing there red-faced and madder than a wet hen. She told us to get her a switch which she was going to use on our butts. Well, I picked up a small stick and handed it to her. This did not sit good with her, telling us that it was not a very cool thing to do. There just happened to be a willow tree nearby, and she went over it and broke off a branch. This was not a small branch either, about 4 feet long. Started swinging that willow branch, using it like a rodeo whip, beating the living tar out of us. We took off running. At this point, I did not know our mom could run so fast. She kept up with us all the way down the road, over five blocks. We were running as fast as we could. She did not miss a lick either, swinging and yelling at us. She whipped us all the way down the road until we got to the Pink Castel. We had some really good welts all over our backsides. Mom was crying, and so were we. Once we got settled and stopped crying, Grandma made us one of the best burgers, and we all sat down and ate together like nothing had happened. Other than that, my brother and I were in pretty good pain!

The Pink Castel was a major focal point for my brother and me as the years passed. Every time Dad would leave, we would come home to Arkansas. And every time we came home, we would stop off to sec

Grandma and Grandpa. And of course, we would always end up in the Pink Castle. No matter what time of the day or night, my grandma would cook us a burger. Also, she would give us a pop and let us pick out one candy from the candy display rack at the counter. It seems my brother would always pick out a pink nut patty. It was a pink patty that had raw peanuts in it. The pink color was entirely made from sugar and colored pink. One of my favorites was the 5th Avenue bar or a Zero bar. Nowadays they are very hard to find in the stores. Also, there was a pinball machine in the Pink Castel, and we would take turns playing it, waiting for the bell to ring so the kids could go back to class before we took off and got bottles for Grandpa. While we were playing pinball, our grandma would bring us a big burger and chips as we were playing. She really took care of her grandkids.

One year, Dad got some leave, and we went home to work on our house. Every time we could, we went home to improve our small house. Our house sits on one acre of land. The house sits right in the middle, too. One year, we added on a new section which is now our big dining room, and we also added on a carport. Two years later, we changed the carport into a giant den with a custom fireplace that Dad designed.

So, the leave Dad got was during the summer months—prime fishing time for Arkansas. Well, Mom wanted to have a very plush yard, and so Dad went to work on making this happen. He went out to the local cotton farms and made some kind of deal with them. It seems that once a cotton boll is removed from its stem, the stems have to be either burned or trucked off to a landfill. Remember, this is the way it was done in the 1960s.

Well, Dad being the deal-maker that he is, made a deal with a farmer to have the stems trucked over to our front yard. He told the farmer he needed about 10 truckloads, too. The farmer agreed, and the next day, all these trucks showed up and started dumping the stems in the yard.

While all this was going on, my brother and I were preparing to go on a fishing trip—famous last words. Dad got us up at the break of dawn and told us to meet him outside. Once we got outside, he handed us a shovel and a rake and said that we had to spread those piles of stems all over the yard.

Two days later, we finished what he asked and again started to plan our fishing trip. Well, not so fast—Dad told us. It seems that while we were spreading the stems, Dad made another deal. He knew a guy who was putting in a new house and moving a lot of dirt. He didn't have anywhere to put it and didn't need it. So this is where Dad came in. He said, as they dig out the dirt, why don't you have them bring it over to my front yard and dump it? I'll need about 10 truckloads to cover my yard.

The next morning, we had 10 loads of dirt where there once were cotton stems—and you guessed it—Dad gave us the shovels and rakes again and said, "I want this dirt spread out all over the front yard."

OMG, spreading dirt is a hard job to do. Our arms and backs were killing us. It took us another three or four days to complete what he wanted. We did finally get to go fishing, and Mom did cook us some of the best steaks with fresh mashed potatoes and some really good homemade gravy with melt-in-your-mouth rolls.

So... if you get your kids to do the work, is it still called child labor?

We did a lot of fishing in Arkansas, too. There is a song out there about running a trot line and skinning a buck and we could do that before we were 10 years old. Our dad taught us how to set up the trot line, how to bait the trot line, and how to run it correctly. Most folks think there is no problem when running a trot line but there is a method. Dad showed us how to tie one end of the line to a tree or stump, run the hooks out, and secure the other end to another tree or stump. The line would be the equivalent of a parachute cord today with about 20 to 30 hooks attached. You had to weigh it down on both ends with big weights or concrete blocks and place the line in an area that boats would not travel through and cut your line. Once the line was secured, you would take your stink bait, livers or dead fish and bait the hooks. But there is a trick to running a trot line. You do not put the line in your hand and pull it. The reason being is if you have fish on the line the fish will pull the line through your hand and you end up with four or five hooks embedded in your hand. How do I know this? The first time I ran the trot line I ended up with at least 4 to 5 hooks in my hand. I thought I had a better and faster way to do it, but there was no way! Dad got a good laugh and showed me how to run it correctly by holding the line with my fingers and gently lifting it up out of the water to where the hooks were hanging.

We also would set out gull nets in the creeks after the Black River flooded. It seems that the Black River would flood every year making the bottom lands around the river the most fertile land around. Our gull nets were about 20 feet long and hung down about 5 feet. We would drag them across the creek, back and forth from side to side and let them set for a few hours. Dad was home on leave for one summer month, and we hooked up with my Uncle Joe, Dad's brother-in-law, and decided to see if we could catch some fish after the water went down. So, we put the old John boat in the back of Big 10, loaded the net, and went off to find a nice creek we could get to on foot. We found a pretty wide creek, about 8 feet wide, and launched the boat and started setting out the net. We set it up and started waiting for it to fill up. Dad put me in the front of the boat so I could run the net. Uncle Joe was in the middle with the cooler. As I got a fish, I would hand it back to him. Dad was paddling in the back of the boat. There were about five or six crosses of the net from bank to bank. As we got into the second pass across the creek, we were getting 4-to-5-pound buffalo fish. The bullalo fish is part of the "Sucker" family of fish, but it is not a Crap.   The bigmouth buffalo fish is typically a brownish-olive color with dusky-colored fins. It can reach lengths of over 4 feet, and it is a very good fish with a delicate flavor. We mostly eat them fried. We were netting fish in lengths of two to three feet, just having a good old time. As we started the third pass, I pulled up a buffalo, about 5 pounds, and it had a chuck out of the belly. My Uncle Joe said that it was caused by a big turtle. He said that the turtles would eat the fish that were stuck in the net, making for an easy meal. On the fourth pass, about halfway across the creek, as I was pulling the net up, I saw a big round log in the net. As I got closer, it came to my attention that it was not a log. Oh no! As I pulled it closer, I noticed that this log had a very ugly face, and its neck stretched out about 8 inches, and it started snapping at me. It was a giant Alligator Snapping Turtle. The Alligator Snapping Turtle has a bite force of 1,000 pounds and can snap a finger off in seconds. I didn't have time to think and once I saw that mouth opened up and coming at me, I jumped out of the boat, walked on water for two steps or so and landed face-first on the bank. As I was getting up, I turned around and my dad and uncle had pissed their pants laughing so hard.

They said I hit the water in two spots and got on the bank faster than a "duck on a June bug." Once they stopped laughing, they pulled this giant of a turtle into the boat. Once they got to the bank and got the turtle out of the boat, it took both of them to move it because of the weight. It took six shots from a .22 caliber to kill it. It was so big that once we got back home, it would not fit into the top of a 55-gallon drum. Of course, Dad and Uncle Joe cleared it and we cooked it up that night for supper. They say that its meat is as light as a walleye, with the dark meat tasting like a grouse and having the taste and texture of tenderloin. All I know is it tasted pretty damn good with some homemade bread! That shell is still at Mom and Dad's house, hanging in the gazebo in the backyard.

Another time that the Black River flooded, we decided to take the old John boat out just to do some cursing in the flooded fields. Dad was showing me the correct way to paddle a boat. As we cursed along in a flooded soybean field, we came across a larger number of Israeli Carp. The Israeli carp has very large scales and a ridge line across the backbone, which is different from the standard carp folks see today. They can weigh in at about 30 pounds. They were just swimming around in a large field, feeding on the soybeans lying on the bottom. So, we decided to head back to the house and see what we could rig up to catch a few of the monsters. We found the old frog gig that one of Dad's friends had made. We figured that, being about 10 feet long, we would have plenty of gigging room once we got onto the fish. We loaded the boat in the truck and off we went to gig us some fish. The water where the fish were was about 4 feet deep and crystal clear. Dad showed us the correct way to gig fish. It seems that Dad knew the correct way to do just about anything. After he gigged a few, he let me try it. There is a trick to it. You never gig a 20-to-30-pound fish with the handle near your face. If you gig the fish, he will swim off with such force that the 10-foot oak pole will hit you upside the face and knock you out of the boat. I reckon I should have watched Dad better, but I got so excited to gig a monster fish that I did not listen to or watch very well how my dad gigged them. I gigged the biggest carp I could see and the next thing I knew, I was in four feet of water with a big red line about 2 inches thick running down one side of my face. Oh yes, that oak handle hit me so hard. I guess you could say it subsequently knocked my block off. Of course, my brother and dad were just dying laughing at me and helping me out of the water. So, from then on, I did it correctly. We feed Dad's chickens the fish we got. This brings me to the altercation I had with a very mean roaster.

Dad keeps one roaster in his pen; he says that if you keep more than one, it will only create problems for the ladies. Sometimes, he would end up with more than one roaster. When this happens, he will replace the old one with the new one. This one roaster he had was a very mean, tough guy. He would peck at your feet and try to get up in your face, pecking at you all the time when you got into the pen. Dad had about enough of his carrying-on and decided that it was time to replace this roaster. He got ahold of him one day when we were out collecting eggs, and he told me to get the axe. Dad wanted me to cut the head off of the roaster. Now I have never done anything like this before. Normally, I would get the 12-gauge shotgun and just blow his head off. But not on this occasion. Dad collected the angry ballistic roaster up by his feet and forced his head onto a tree stump, which was in the chicken pen. We used to cut the heads of the chickens before we ate them and he told me to cut his head off. Well, everyone has heard about the chicken with his head cut off. Guess what? It works the same way for roasters! I took that axe and, with all my strength, swung down and separated his head from his body. Now this roaster was bleeding like a stuck pig and was chasing me all over the pen, blood flying everywhere. My little brother and my dad were just beside themselves laughing at me. There was blood all over me and the pen and I was scared to death! Of course, that roaster made some great chicken and dumplings that night!

Near our little town in Arkansas, there were numerous rivers, lakes, and creeks. You could not drive five miles without crossing some kind of creek bridge. Also, our town had countless railroad bridges that would travel across the creeks and rivers. Not far from our home was such a creek. It not only had a bridge for vehicles but a railroad bridge that crossed it. On numerous occasions, we would get our gear together, load it on our bikes, and ride down the road to the creek. The creek was called Village Creek and was only about five miles away from the house. Along the way, we would stop at the local filling station and get a pop, some chips, and a bunch of crickets. Crickets were our choice of bait for fishing this creek. We had picked this day to go fishing because Black River had flooded a few days again and the sun was out, making for a very good day for catching fishing. We got to the creek and unloaded our poles and tackle boxes and walked down the bank. The water was up and moving pretty fast, which made fishing so much better. We decided to fish under the road bridge next to the railroad bridge to see if we could catch some bluegills and maybe even some nice catfish. The creek produced some very nice-sized fish last time we fished it, and a blue gill is a nice fish to eat. After a few hours, we decided to cross over and fish from the railroad bridge. We got our gear and walked out to the middle of the track crossing and sat down just about in the middle of the bridge and started fishing. We must have caught about five or six nice fish just having a good time, drinking our pop and eating our chips. As we were fishing, we heard a train whistle blowing. We live in an area where you hear a train whistle everyday so we did not think much of it. We kept fishing, and all of a sudden, the bridge started shanking and vibrating in a crazy way. We looked down the tracks and this big long train was just about on us. We picked up our poles and tackle boxes and started running as fast as we could. We had removed our shoes, so we were barefooted and those railroad ties had some pretty sharp slivers on them, but we didn't think of anything but making it to the bank so we could get out of the way of that train. At one point, my brother and I realized we would not make it to the end. So, we decided that the only way we were going to live was to jump off the bridge into the roaring, brown fast flowing water below. The drop to the water was about six to seven feet and we jumped off the bridge, one on each side. Once we hit the water, our tackle boxes and poles went everywhere, mainly traveling fast down the creek. We swam as hard as we could against the current and made it to the bank. We pulled ourselves up the muddy bank, all out of breath and just looked at each other, realizing that we just saved our lives by jumping off the bridge. Then we realized that we had just lost all our fishing gear. After a few seconds, we busted out laughing and wanted to jump off that bridge again.

We had many learning experiences fishing. This time, my dad and Uncle Joe, along with my brother and me, went fishing in Lake Charles. Lake Charles is a pretty good-sized lake that runs along the Black River. It was pretty hot in Arkansas that week but Dad and Uncle Joe wanted to go fishing. They were fishing in a little cove and hooked into something pretty big. My brother and I were fishing a little further away but could see they had hooked up. We'll see all the boats that were fishing around them start moving over to see what they had hooked into and maybe horned in on their fishing spot. Dad was fighting this fish for about half an hour and all of a sudden, he pulled it up and it was a 5-foot water moccasin. Once the snake hit the top of the water, all the boats took off like a bat out of hell. My dad and Uncle just sat there laughing.

We learned a lot about fishing and how to catch fish in different ways. One of those ways was to catch fish on a rig called a yoyo. A yoyo rig is what it sounds like. It is a metal spring-loaded yoyo with a hoke and sinker attached to it. You bait the hook and set the lock that holds the bait at a certain deep. You attach it to a tree or over handling branch and sit and wait for the fish. After a while, you can hear the yoyo go off as the lock disengages from the spring and hooks the fish. Once they bite, the spring reals them. As you are riding along in the boat, nothing sounds better than hearing a yoyo go off and see the fish flipping in the water. We did this in Spring River and we would catch all kinds of fish. Another way to fish is to jug fish. The way to jug fish is to get a gallon milk jug and tie a line through the handle with a hook and sinker, bait the hook and you drop it in the river. You would take your boat upstream, throw the jugs in the water and watch them slowly drift downstream. After your jugs are out you would go check your yoyos, collect the fish from the yoyos and then start looking for your jugs. Fishing in Spring River, you never know what kind of fish you will catch. Spring River has anything from trout walleye, to largemouth and smallmouth bass, to channel catfish. If a jug is floating and moving around, you've got a fish. One of the most exciting things in the world is pulling up the line from a jug to see what is on the other end. We have caught tons of fish using both methods, and it never gets old pulling up a 5 or 6-pound channel cat on a jug line!

Dad always showed us how to fish and hunt. When we would go hunting, Dad would take the 22 rifle and give us the 12-gage shotgun and the 410-shot gun and we would go bird, rabbit and squirrel hunting depending on what time of year and season it was. Normally, Dad would use the 22 rifle, and I would get the 12-gauge, and my little brother would get the 410. Dad would use the 22 rifle if we boys missed. But whatever you shot, you had to clean it. Again, this is another teaching lesson for us on how to clean the different animals we have. We were raised on a wild game. And my mom could cook up some delicious fried squirrel with homemade mashed potatoes, brown gravy, and homemade rolls. One of the worst things in the world, though was to start chowing down on that fried squirrel and bite into one of those pellets from the shotgun shells. Oh man, that hurts so bad!!! If you bit down too hard you can break a tooth. Thank goodness we never did.

One experience that my brother and I will never forget is one where we went up to see Uncle Joe, who lived in Mariana, Arkansas. It was about a three-hour drive from home, and my dad and Uncle Joe were pretty inseparable. Uncle Joe was married to Dad's little sister and they were closer than two peas in a pod. You see, in the little town of Mariana, Arkansas, our Uncle Joe was the mayor, justice of the piece, and game warden. He was also the publisher and printer of the city paper. Every time we would get together with Uncle Joe, we would ask him to count to ten with his fingers. He would count out loud using his fingers. The only problem was he never could get to ten using his fingers. The worst he did one year was eight and a half. Reason for the eighth and a half was that he kept losing fingers. One year, he lost half of a finger and the next year, he lost a whole finger in the printing press! But our experience began with a swimming trip to Bear Lake. My brother and I along with our cosine Jerry, took out the little boat so that we could get to the diving pier that was out in the middle of the lake. We had been out there for hours, drinking beers and swimming. We all decided it was time to head in and I am not sure how it happened; it might have to do with the beers, but we tipped the boat over. We got out OK but the boat, well it is at the bottom of the lake now. We swam to the bank and went up to the old cabin that was nearby. The cabin belonged to Grand Jodie, Uncle Joe's dad. Grand Jodie must have been in his late 70s at the time but still could outhunt anyone in the county. As we started up the driveway to the cabin we noticed a giant car, all white, and a convertible as well. We went in the door and sitting at the dining room table was our dad, Uncle Joe, Grand Jodie and Conway Twitty. They were playing penny poker. Now, we three knew who Conway Twitty was but could not believe our eyes that he was sitting at the table with our dads! As we looked around the cabin, we saw photos of Grand Jodie and

Conway and all kinds of record covers all over the walls. All of a sudden, a very good-looking young woman came up to us and started talking. After a while, she asked us if we wanted to see Conway's car. The white Lincoln Continental outside belonged to Conway. We went outside with her and started looking around the car. It was all leather and had all these gages and lights everywhere. She reached in and turned it on and all the lights and whistles you can think of started up. That car was the most beautiful thing we had ever seen! After a while, she told us a story of how Conway Twitty got his name. It seems that Conway Twitty, whose real name was Harold Lloyd Jenkins, was from Mississippi and he wanted a stage name. So, he put a map of Arkansas on a table, closed his eyes and put his finger on the map. He had placed his finger on Conway Arkansas. As he was looking at the map, he noticed Twitty, Texas, nearby, and that is how history remembers him. I came to find out that Conway Twitty is related to my cousin Jerry by marriage. Jerry's wife is Conway Twitty's cousin.

Chapter 3
France

In the year of our Lord, 1968, my mom gave birth to our little sister. Within the next year, Dad had a posting in France. We found out later that we were the last US troops to be stationed in France. We lived on the economy, which means we did not live on base in base housing to start, which means there were no government quarters available on base for us to live at the time. The place we lived was in a five-story apartment building not too far from where Dad was stationed. We were on the fourth floor, and the stairs were very steep, which was a good climb. Our place was very small for our family of five but my brother and I were in "hog heaven" because we were in a foreign country and we just wanted to explore everything. We can remember Dad getting on to us about hauling coal up the four flights of stairs to our apartment. Our little apartment was heated by a coal-burning fireplace, and the coal for the building was loaded into the basement. You had to get your own coal if you wanted to keep warm. The building manager gave all the tenants these 10-gallon bags made of burlap. You had to walk downstairs to the basement and bring the coal up. You should fill the bags to the top and carry them upstairs. But my brother and I felt that the bags were too heavily filled to get up the stairs. For the first few loads we took upstairs, we only filled the bags half full, thinking we could get away with a few trips, but Dad was wise to our ways. After the second set of slaps across the face, a reminder to "do the job correctly the first time," which was my dad's favorite saying, we filled them to the max. These bags had to weigh in excess of 30 pounds and again Dad would say, "Hard work builds character"! Our apartment was heated with coal so if Mom wanted the house warmer, we had to make more trips each week.

Also, during this time, we did some exploring. We found that there was a small river not too far from our apartment and we went off to see what was going on. On this particular day, we found a small boat tied up on the river. Well, we decided that this was a time to take a little boat ride. My brother and I got in and went on a little trip up and down the river. But when we came back to tie it off, the owner was there. And so was Dad. This is the proper time to let you know that Dad had a system. He would whistle for use when to come home. If he had to whistle three times and you did not come home, he would take off looking for you. So, Dad had whistled three times, and now he is waiting for us at the bank with the owner. To this day, I do not know how Dad knew where to find us, but he did. The owner of the boat was pissed off at us for taking his boat without permission. So, once we got on the bank, he was right in our face. He was yelling in French at us and started to hit me, me being the largest of the two. He pulled his arm back and started forward but Dad caught his arm and hit him right in the face, knocking him down. He did not know much French but as he was standing over him, he told him, "I am the only one that will hit my boys," to where he proceeded to wipe our butts. I guess this would be the beginning of the Austini Averages.

On another occasion again at our apartment, there was an old French made truck that would come around and sell candies to the kids around the apartment area. The truck had one side that had a sliding door that went halfway up the side and once the door was opened, all the candies were on display. Well, my brother and I did not have much money, so when the truck came around, I, being the oldest and the tallest, would talk to the driver while my brother sneaked candies and put them in his pocket. These candies were the best ever. The best of the bunch were these seashell halves filled with hard candy. You could put the candy in your mouth and suck and lick the candy out of the shell all day. Most excellent candy! We got away with this for a few days till the driver got wise. Once he caught my brother in the act, we just took off running so fast that no one would catch us. But somehow, Dad found out. When we came home, he asked us if we liked candy. He asked us if we knew that stealing is not what young boys should do. This led to our butts getting whipped for stealing. Another lesson from our dad is that stealing is not the way to move forward with our lives. He said, "Hard work and perspiration build character".

While waiting for government housing, we moved a few times within the little town of Verdun, France. One moment in our lives that we will never forget is when our little sister, who was about nine months old at the time, reached into the small

European-type refrigerator and dropped a large jar of pickles onto her foot. The floor was marble and very slick. This may not sound like much, but the small refrigerator was on a two-foot-high ledge and she had just started walking. She reached into the refrigerator and pulled the jar out and dropped it onto her second toe on her right foot and just smashed it off. The sound that she made, my brother and I will never forget and neither will my mom and dad. They took our sister to the hospital with the smashed toe. The doctor sowed it back on and said someday, my little sister might want to wear sandals.

One of the many good things about being in the military, other than seeing the world is that the military has all kinds of sports programs for children of all ages. This is where my brother and I started our sports careers. The base dad was stationed at had sports programs from football, baseball, and basketball for all age groups. My brother and I can remember playing tackle football. My brother, who is 13 months younger than I am, was put on the Dallas Cowboys, and I was in the next age group, so I was put on the Green Bay Packers. And to this day I am a Green Bay Packers fan. I can still remember the time we had a game, and the coach called my name and told me to get in there and block that punt. I was playing inside down linemen at the time. As the play started, I broke through the line, and the punter kicked me with the ball right in my stomach. Oh, golly my stomach hurt for days but we won the game.

Another sport my brother and I played was baseball. The sports program at the base was always looking for coaches, and Dad stepped up to the plate and volunteered to coach a youth baseball team. During tryouts, Dad of course, picked his own sons for his team and we became the St. Louis Cardinals. There were many teaching lessons during this time frame but a few really stuck with me and I used them when I started coaching baseball. The lesson was that Dad would pitch to the batter, but the balls would be partly painted red or blue. When he pitched it to you, he would ask you what color the ball was. This will teach the batter to keep his eye on the ball. The other thing he would teach us was that, as a pitcher, you need to follow through when you release the ball. So, he would put a quarter in front of the pitcher's mound, close enough to reach. As the pitcher would get through his wind-up and release the ball, he would have to pick up the quarter. This would instill in the pitch to follow through the whole pitching process.

My dad coached us boys for a few years, but due to his work, he didn't have the time to keep coaching. Still, my brother and I continued to play ball. One year, we played on a very good baseball team. Not many folks would come to our games, but that didn't stop us from playing as hard as we could.

We were late into a game once when Dad was able to come watch. He sat right in the middle of the bleachers with his cooler filled with a six-pack of beer and watched us play. At that time, I was playing third base and my brother was on second. One of the other team's players connected with a pitch and hit a triple, ending up on third base.

Now, this boy had been giving me crap all game. He knew I was a pretty good ball player, and he thought he was so much better than me. While standing on third, he kept getting in my face and yelling some really colorful adjectives at me. My dad was watching and could tell I was getting pretty mad.

All of a sudden, he yells out, "If you want some, get some!" So, I called time out, walked right over to that boy standing on third, and hit him in the mouth as hard as I could. Oh yes, I knocked him out cold!

Now, third base was right next to the other team's dugout, and as soon as that boy hit the ground, their whole team came running out to get me. Well, my little brother was on second base, and before they could reach me, he was there. And just like that, we started fighting the whole team.

Yes, we both got kicked out of the game—but you know what? The stands were packed with fans every game after that. Everyone wanted to see if there'd be another fight!

My brother and I have had many adventures together, and some of them did not end with a butte wiping. One time, our dad took us fishing. We both loved to fish, coming from a small town in the south where you were either hunting or fishing in the water. But this was not your ordinary fishing trip. First off, we had to go to a US Army German post a few hours away and it had a monster lake. It seemed to be a monster of a lake to us because we were small, but it was big. They were killing the lake and draining it to clean it up and restock it so that those stations at the Army Post could enjoy some rest and relaxation, R and

R. As the water got lower and lower my brother and I had never seen so many fish in our whole lives. Folks were netting them like crazy and putting them into these metal Army-issue ice coolers, which were about three feet by two feet with a removal cover (they are called ice chests now). We must have eaten fish for days, but we never got tired of it. Dad was a young Segert, stationed overseas with a wife and three young children, so we did not have a lot of money, so he took advantage of every opportunity to feed us for free!

While we were in Verdun France, we did some sightseeing. My brother and I can still remember many of those trips. During WWI, Germany staged a major offensive in France. The Battle of Verdun went down in history as one of the longest and bloodiest battles of the war. Along the River Meuse, the French have set up a memorial to those who had fought and died during the battle. The trenches used during the battle are still there. The French enclosed the trenches with blueish-colored glass and put them on display so you could see what it was like back in the battle. We saw skeletons with weapons and shells all lying about in the trench. It just stuck with us to this day the way it all looked.

While traveling to France, we got into the habit of eating warm, fresh French bread and freshly made chocolate. We would load up in the big station wagon to go on another drive through the countryside, and before we started on our trip, we would find a local bakery and time it so that the bread would be coming out of the oven and buy it still steaming hot. Then, we would find a local chocolate shop and get our chocolate. That was some pretty good eating back in the day, and it was cheap, too.

Time moved on, and we both remember going to the school on the hill in France, located at the Army Post. Our school was on top of a big hill, which was also the Medical Group building. Our classrooms were in one of the big wings of the medical faculty. But what my brother and I really remember the most was that the hill was surrounded by a nine-hole golf course. And at that golf course was this snack bar. They made the best hamburgers you have ever put a lip around. They would put these grilled onions on the burgers and just take them over the top. Also, we would get a 7up with our burger. They came in a light green bottle, old-school type, which brings me to the next adventure we did as boys. We would eat our burgers and then down the whole bottle of 7up. Doesn't sound that bad but have you ever tried to drink a whole bottle of 7up without it coming out of your nose or throwing up? Well, we tried and every time, we made a mess!

We finally moved into base housing. This would become a game changer for both of us that would last us for the rest of our lives. We had a bunch of buddies that we hung out with and we would go to the back part of the base and build these big forts. This area of the base was somehow involved in WWI, and there were many foxholes and some small underground tunnels. We worked on the tunnels, and we used cardboard boxes that the movers had left to build our above-ground fortress. We pretended to be at war and had battles. This is where life-changing actions happen. Our dad smoked a lot. We thought that it looked so cool that we would steal smoke from our dad and bring it out to our fort and smoke like grownups. It seems that every Army soldier knows how to play penuckle. It was a game that brought the local folks together and fellowship. Most of the time, the games would last until late into the night. Mom and Dad would play every weekend with their favorite couple. Well, one night, while playing cards, the couple they played with started telling a story of how their boys were stealing cigarettes from their house and going out to smoke them with their buddies. They caught their son doing it and asked if Dad knew that his boys were stealing smokes from him as well. Well, this was the first time Dad had heard of this. This really pissed our Dad off too, not knowing that his boys were stealing from him. So, he told the couple right then and there that the game was over and that they needed to leave. After they left, my dad called us boys into the dining room and had us sit across from him at the dining room table. He took out a pack of cigarettes and fired up a smoke. He asked if we wanted one. Of course, we said yes; to smoke with our dad would be cool. Well, that was a very big mistake. We puffed on our cigarettes for a while, and Dad said that we were doing it wrong. He said this is the way you do it. He drew in a big draw off of his cigarette and held it in before he exhaled. He told us to do it, so we took a big draw. Again, we thought it would be so cool to smoke with our dad. After a few puffs, he said do not let it out; you need to swallow the smoke. This was not cool. He made us keep going this for a whole pack of cigarettes and we both were turning green. I was the first one to run to the bathroom and puke, spewing my guts out.

It was not long before my brother showed up and spewed all over the bathroom. We never smoked again and to this day, we don't smoke and even the smell turns our stomachs. Life lesson learned again from our dear old dad.

Chapter 4
Washington DC Fall Church Virginia

As time passed, the Vietnam War was going strong. Our dad was part of the infantry, and he was finally called to deploy to Vietnam. But before he could go, he had to learn the Vietnamese language. It seemed that he was going to go over there as an advisor to help the Vietnamese people fight. So, we got an assignment in Washington, D.C., so Dad could go to the Vietnamese Language School. We lived outside of DC in Falls Church, Virginia. We lived in an apartment complex about a 45-minute drive from where Dad's school was. This was just another adventure for my brother and I. One of the best things to do in Falls Church was to go dumpster diving. Now, this was in the early to mid-60s, before folks knew what dumpster diving was. Near our apartment complex, there were these 40-story offices and in the rear of the building, there were these monster dumpsters. Within a block, there must have been ten dumpsters. We just started jumping into them but then discovered that folks would throw away all kinds of things. We found money, tools, jewelry and all kinds of office equipment. But on some occasions, we would jump into a big dumpster and end up in someone's old lunch. Not cool and very stinky. One man's trash is another man's treasure most of the time! We would bring this stuff home and Dad would get pissed at us for bringing home all that trash and would make us throw it away, but we never told him about the valuables.

Also, it snows in Falls Church. It was not uncommon for us to sleep with our windows open and on many occasions, we would wake up to two feet of snow in our bedroom.

Another weird thing that happened during our stay in Virginia was that one day, we noticed that Dad had a blister, which was a little bigger than a fifth cent piece, on the left lower arm. We asked Dad how he got the bad burn on his area. He told us that he was "volunteer told"; this is where, in the military, you get volunteercd to do something, and if you don't, you face the consequence of not volunteering, whether you like it or not. They injected into his arm an experimental drug the military was testing. Later, we found out that the drug was for some kind of blistering agent that the North Vietnamese were thinking about using against the American forces that the North Vietnamese army got from the Russians. To this day, he has a bad scar on his arm. My brother and I, to this day, remember it and how bad it hurt our dad for days! The good thing that came from all this is that our forces were prepared for this in Vietnam. Thank God Dad never came into contact with the blister agent.

Chapter 5
Ft Leonard Missouri

D ad was sent to Ft Leonard Wood, Missouri, after he survived his 18 months in the Vietnam War. Again, it was another adventure for me and my brother. Dad would take us hunting back up in the woods around the post. This was another learning experience for both of us. He taught us that you never point a gun at anything you do not want to kill. This was after he returned from Vietnam, too. In our area of the post, there were these rolling hills covered in thick green grass. This is where my brother and I learned to slide down the hills on cardboard. But again, it is another learning expertise that would cost me. At our house, the backyard was the best sledding hill in the world. During the snowing months of the year, we would get our sleds and fly down the hill. The only problem is that once you get up to speed, there are large trees near the bottom. My brother and I would see who could make it through the trees. Sometimes you would make it, but most of the time, you would end up upside down against a big oak tree. Never broke a bone or got knocked out, but sure did hurt when you hit one. Which brings me to what the sledding hill cost me. During the summer months, the hill was transformed into a thick, lush green grass hill. You would get a cardboard box and make it into a sled. Cardboard was so easy to get from all the folks moving in and out of base housing. Get a good run at the top and jump onto the cardboard. Then, go for an amazing ride right into the large trees at the bottom, avoiding them as best you can. One nice warm day equipped with my cardboard sled, in shorts and a tee shirt, my brother and I decided to go for a ride. There are some things folks need to know about sliding on grass with a flat piece of cardboard. If you slide down a hill, you need to put your hands and fingers inside of the cardboard once you start sliding. You need to bend the cardboard edges up and hold on to them as you slide cause if you don't, you could break your fingers. Well, I was flying down the hill doing just fine till the cardboard shifted. I must have turned the edges down just enough to where my right hand came between the grass and the cardboard. I flipped over a few hundred times and came to rest at the bottom of the hill against a pretty good-sized oak tree. I mentally took note of my body parts, moving my legs and arms to make sure I didn't break anything. Well, I was not so lucky after looking at my right hand. My right pink was pointing in the wrong direction. After we had left the doctor's office, it seemed that I had broken and dislocated in two places my left pinky finger. To this day, it looks pretty funny!

Every Thursday in Ft Leonard Wood was Pizza and Pepsi night at the Austin house. During these times, there were not many takeout pizza places, and we could not afford the cost of takeout either. No, mom would make Chef Boyardee pizza out of the box and a Pepsi, always in a bottle. My brother and I can remember that mom's favorite soda was a Pepsi, which was the only pop dad would buy! Mom would bake a box of Chef Boyardee Pizza up, open some pops and we would sit and watch Walter Cronkite doing the news on our small television and then watch the Thursday night movie. This became a tradition for many years.

There were many more experiences during his time. Like the first time, my dad busted his ass in front of us boys. My brother had developed into a wizard on a skateboard, and one day, he was outside doing his thing on the board and Dad said I bet I can do that. So, we just sat and watched as Dad got on the board for about two seconds and fell and busted his butt but good. Of course, we both laughed our ass off and then we got a whipping for laughing at him, but it was well worth it! You know he never got on another skateboard again.

There were a lot of wooded areas around our house on the post. We had a pretty good walk through them to get to school. A lot of the kids at school said that the woods were filled with "walking stick" bugs. They were talking about the Praying Mantis. They told us that you had to really look hard at the small branches in the woods to see them. This was the first time we both saw one. The images still stick in our heads as we watched this "walking stick" eat another one, just bite the head right off. Thought it was so cool! I came to find out that the female ate the male.

Our school has decided to hold a Smiling Contest. I'm not sure why the two of us remember this, but back in the day, they had all kinds of crazy contests in school. My brother, a total smart-ass, thought it would be fun to enter the contest. They told him he had to smile and that they wanted to take pictures of his teeth.

After three or four tries, he just kept cutting up and not doing what they wanted, they finally got a good picture and wouldn't you know it, he won the contest. It was the first thing he ever won! We still have the certificate on the wall at Mom and Dad's house in Arkansas.

Chapter 6
New Jersey
Ft. Dix

Time moved on for my brother and me. We moved to Ft Dix NJ the next year where Dad had to carry the three-star general's flag. At the time, I did not understand why Dad had to carry a flag and he had to wear a silver helmet, a big silver belt buckle, and white gloves. Well, little did my brother and I know that we were about to become the best silver and boot shiners in the world. Dad would give us either the helmet and a boot or the buckle and a boot and tell us to shine them and show him once we thought we were done. At first, we both just put polish on them and gave them back to him. Not cool! He would look at them and backhand us across the room and say you will make these sparkle or I will whip your butts every time you bring them back to me. You may not think that this is a life lesson, but when I enlisted in the USAF, I made money on shinning boots and polishing brass belt buckles. Also, at Fort Dix I had my first concession. All the base houses were above ground, four feet above ground. All the pipes and wires would run under the houses. Well, my brother and I would build our forts under the houses. And remember the three whistles that Dad would do for us to come home. Well, one day he whistled and we both got up from our fort and took off. Will, at that time, I was a little over four feet tall, so I raised up and started off. There was this big pipe right at the end of the house and I hit it head-on, clipping the top of my head and knocking me out. Of course, my brother just stood there laughing until I came too. After a trip to the hospital and five stitches, I whipped my brother's butt.

While we were stationed at Ft. Dix, Dad became a Drill Instructor (DI). Being a Drill Instructor is one of the most demanding assignments a soldier can have. An instructor is a symbol of excellence, an expert in warrior tasks and battle drills, and a true professional. Once you become an instructor, you take on the responsibility of coaching and mentoring everyone assigned to you. You are responsible for transforming a civilian volunteer into a combat-ready soldier. You become everything that civilian knows about the military, through your teaching. And my dad was all of this and more.

My mom put so much starch in his uniforms that they could stand on their own in a corner. He was on call 24/7, always ready to instruct civilians on the military way of doing things. He also carried his own version of a swagger stick. The military swagger stick is a short staff carried by a Drill Instructor—it's a symbol of authority and leadership and a part of military history.

Dad's swagger stick was made of steel. The tip of the stick was the head of a .30-06 shell, and the butt was the cartridge. In between was a shiny steel shaft. It weighed about five pounds, and Dad used it every day with his troops. He told us he chose the .30-06 rifle cartridge because, at that time, it was one of the greatest rifle cartridges ever designed.

On several occasions, Dad would take us boys with him during some of his inspections. Not really knowing what was going on, my brother and I would tag along with him to the barracks. To our surprise, he'd begin a full inspection of his troops' beds and personal space— including each soldier's footlocker.

When we arrived, the troops would be standing at attention near their footlockers. This is where the fun started for Dad. He would bounce a quarter off their bunks, and if the quarter didn't bounce, he'd flip the bed over using only his swagger stick. If a footlocker wasn't in the correct order, he'd flip that too—with the same stick.

Once his inspection was done, he'd turn to us boys and tell us to "give him 50," meaning we had to do 50 push-ups. Once we finished, he'd turn to his troops and say, "If my boys can do 50, then you all can do 100."

Why am I telling this story? Because many years later—right up until we were in high school—Dad would still make us boys make our beds with hospital corners, or military corners as they're sometimes called. And yep, he'd bounce a quarter off our beds. And if it didn't bounce? You guessed it—we had to redo it.

I guess you could say Dad brought his work home with him... and us boys were right in the middle of it!

When we were in Fort Dix, I can remember we only drank Pepsi because this was Mom's favorite soda, so Dad only bought Pepsi. My brother and I would go out and collect as many Pepsi bottles as we could for some cash. They were giving 5 cents a bottle at the time. Once, while collecting bottles, we noticed that Pepsi was giving away a Pepsi Racing Jacket to one lucky person. So, my brother and I entered as many times as we could. Sure enough, my little brother won the jacket. After he got the jacket, he would never take it off and he wore it everywhere he went.

Chapter 7
Texas San Antonio

After Fort Dix, Dad got an assignment, and we moved to Fort Sam Houston, Texas. Here is where my brother and I started high school. So, we both were at that age when the world was our oyster, and we were trying to figure out how to get the pearl. At this time in our lives, we were into riding bikes and building tents. We made friends with a few other boys, and we got pretty good at making a square tent. A square tent is where you would button up to pup tents and, using the tent poles, make a tent that was a square. It seemed that all of our fathers had come back from the war with tents and sleeping bags so we would spend most of our weekends sleeping in the back yard. So, our sleeping outside would bring my brother and I to a new chapter in our "raising holy hell" time frame in our lives.

In Texas, the mosquitoes can get pretty bad. The base used to run mosquito fog trucks through the neighborhoods. These were trucks that sprayed insecticides to kill bugs. They would drive up and down the streets of base housing, spraying thick fog everywhere.

So, my brother and I decided it would be cool to ride our bikes behind the truck in the fog. We did this for two reasons: one, it was just fun riding in the fog, and two, we figured if we got sprayed by the insecticide, the mosquitoes wouldn't get us. Today, if you did that, they'd probably throw you in a mental ward and check for brain damage due to the insecticide! Maybe that's why, to this day, my brother and I both have breathing problems!

Another memory from Ft. Sam Houston was that, being a training base, it had one of the longest and biggest obstacle courses you'd ever see. My brother and I used to run the course for fun, just to see who could get the best time.

Obstacle courses typically consist of 10 to 12 obstacles lined up in a row, and a person or team runs the course to test their physical fitness. Some of the obstacles included beams about 18 feet long that stretched over deep ditches, climbing a rope 1½ inches thick and about 12 feet high, cargo nets about 10 feet high, 4-foot walls to climb or jump over, posts spaced out that you had to run around, logs to weave through—going over and under them—running up one incline and down the other, and swinging on a rope over water.

We would spend days doing this. Oh yes, we'd sleep like babies at night, too.

Looking back, this turned out to be more important than we ever realized—because when we both enlisted in the military, we had to do these same kinds of courses. Thanks to our experience at Ft. Sam, we were at the top of our class.

Ft. Sam Houston was one of the largest training bases in the United States, so there were many acres to cover and this was like Hog Heaven to us boys. There was a large abundant number of red-winged dove and red fox squirrels throughout the base. As boys from the south, we were raised on wild game including doves and squirrels. At Ft. Sam Houston Dad got us both pellet guns. It was a pump pellet gun, and the more you pumped it, the stronger the pellet would travel. My brother and I would take the gun and shoot at anything that would move. We noticed all the game in the area so we started to hunt them. Many times, we would bring home a few doves. Dad would show us how to clean them and mom would show us how to cook them. We also hunted for the red fox squirrels. These squirrels became a challenge but we were both up to the challenge and hunted them a lot. Both of us loved to eat the game, and Dad kept enforcing the rule that "you eat what you kill," so you never overdid it. The problem was that the SPs (Security Police) on base arrested us with two squirrels and four doves in our pouch. They said we were hunting illegally on government property and that we did not have a state license to hunt small game. They also said that we were discharging a weapon without a license too. They called our dad and he came down to the police station. He started arguing with the police about how we had a pellet gun, not a real gun, and that on-base taking the game was illegal. He was called into the base commander's office, at his request, to get to the bottom of the problem. My brother and I were also called into the office. This is where my brother and I could not believe how our dad went off on the Commander. Now, this is what they called a Full Bird Colonel, in charge of the whole base. He was yelling and standing over him.

This is where Dad told us to go outside. We went and sat outside the office but we could still hear Dad going off on the commander. We could not believe what Dad was doing. Soon, he came out and told us to go because we were done. To this day, we joke around about how it all ended, but we kept on hunting and Dad did not lose a stripe.

Another avenge was when we would make those square tents in the backyard and sleep out on weekends with our buddies. We would ride our bike around the base at night and go to the golf course and get the flags from the holes and do some jousting on the golf course. After a few nights of that, we moved on to the laundry rooms that the barrack had in them. These were very large washers and dryer, big enough to get into, so we did. We would get into them and turn them on and just go round and round till we spewed our guts out. We then found out that there was a nurse's barrack on base. We took off on our bikes one night and found all their clothes on the laundry lines too, where we took all their underwear and spread it all over the base. Dad got word of this wonderful adventure and sat us down and asked us if we knew anything about it. Of course, we told him we didn't know what he was talking about till he showed us some female underwear which led to a large ass whipping.

Also, vending machines were in all the barracks. There would be a vending machine for sodas and another one for snacks. But some had a vending machine for beer. Again, we would put up our square tent and sleep outside on the weekends. We would go into the barracks at night and play pool all night and we would get hungry and thirsty. We figured out, after a few tries with the beer vending machine, that if you put your quarter into the machine and hit the button the little door would open at the bottom and the beer would fall out. Well, it did not take long for us to figure out if you unplug the beer machine at the right time that the door would stay open. Once the door was opened, you could pull all the beer out of the machine by just putting your arm up into the door and getting them. It did not matter what beer we got because it was all free. The snack machine was a little different. We could not figure out how to get the snacks out without breaking the glass so we just picked up the machine and turned it upside down and dumped it out! Of course, the Army troops got blamed for all the missing food and beer but after a while, we got bored with that. Of course, my dad, who was always in the know, heard about machines in the barrack that would be empty without any money when the vendors came to refill them. He asked my brother and me about it and of course, we said we didn't know anything about it, which led to another ass-whipping. He said he whipped our ass because he just knew we were involved.

Also, in Texas, it gets pretty hot and we used to ride our bikes in our shorts and tee shirts. In the back of the base, there were these motorcycle courses with big hills and 90-degree curves. So, we took our bikes and rode the course. We would take the hills and do tricks till one day we both took a hill and got some big air and did our tricks but we missed the landing. We ended up in a cactus plant, a prickly pear patch they call it. Now these stickers are like porcupine quills, very hard to get out of your skin. We had stickers all over our backs and butts and of course, we just sat there and laughed both trying to remove them, little devils. On another occasion, we were hitting the hills hard and getting some good air and my little brother came down hard on the ground and gave out a loud yell. What happened was when he fell the jack stand on his bike entered his left calf right in the meaty part. He was just lying there with the jack stand in his leg and bleeding like a stuck pig. We were so far away from anyone, and I was about to get on my bike and get help when this guy in a pickup truck came over and asked what was going on. He said he heard a loud scream and was wondering what happened. One look at my little brother and he knew. There was blood everywhere. He removed the kickstand from his calf, wrapped the wound and loaded us up in the bed of the truck and took off for the hospital. They called Dad and he came out. As the doctor got ready to take care of my brother, he asked me if I wanted to watch. Hell, yes, I wanted to watch. There was a perfect hole about the size of a nickel in his calf. You could see the muscle and all the inside stuff and they started pulling out the damaged parts and put about eight stitches in his calf. That was so cool, other than my brother screaming in pain.

Again, in Texas, it gets very dry. Dad talked to us about how dangerous a fire could be or how fireworks could start a fire due to the dry areas on the base. He told us stories of how an M80 could blow your hand off. He told us stories about how he and his brother-in-law (Uncle Joe) would go fishing with them. They would drop them off behind the boat motor and let them sink and once they blew up the fish would be knocked out and you could just scoop them up in a net, about how one time the M80 just kept spinning around behind the boat by the motor and blew up the whole motor.

After these stories, we thought that Dad had some M80s somewhere in the house. So, we being so curious, we went through Mom and Dad's dresser draws and we found a whole bunch of M80s. We took a few and off we went on our bikes to the back side of the base. On back side of the base, there was an obstacle course set up as a training area for troops to go through. On the back side of the base was also some of the biggest red ant hills and we were on a mission to blow them up using the M80s. We had blown up about two or three of them and it was getting late and we were too far to hear Dad's whistle, so we decided to blow up this one monster hill over by the trees. Oh, it blew up alright and caught the trees on fire. My brother and I started throwing dirt on the fire but it got into the grass and just took off. So, we got on our bikes and headed home. Now, we always had dinner together; we all would sit down at the dinner table and talk about our day. Our TV would always be on and on this day, the news reported how one-third of the base near the training area had caught fire and burned down that day near the obstacle course. Dad asked us if we knew anything about it and my brother and I did not skip a beat and said that this was the first time we had heard about it. Dad put two and two together and figured his boys had to be involved and found some M80s missing from his bottom dresser draw and began to whip our ass. I should mention at this time, my little brother was so cocky that he thought that if he put some comic books in his pants, that dad would not see them and it would not hurt so much. No, this didn't work and it only made it worse for us both. See, Dad started whipping our asses and saw the comic books and told us to both take our pants down, including our underwear, to where he started whipping our ass bare butt. From then on, our ass whipping was without pants.

Another adventure my brother and I had in the great state of Texas was with a shotgun and some prairie dogs with a few cows thrown in. My dad had and old retired Army friend who owned about 1500 archers of land in the panhandle of Texas. He invited us to his ranch whenever we were back in the States. He had over 300 head of cattle on his ranch, raising them for meat. So, one hot summer, we took him up on his invitation and took a trip from Ft. Sam Houston to his ranch, which was about a four-hour drive. When we got there, he laid out his plan. It seems that the prairie dogs had taken over a great portion of his land and he needed to remove them. His cows were stepping in the holes and breaking their legs, which meant he was losing a lot of money. He wanted us to load up in his big Dodge pickup truck and take some guns along with a cooler full of refreshments and go hunt us a shit load of prairie dogs. So, off we went. He would do the driving, Dad would be his navigator and beer supplier, and my brother and I would be the shooters from the back of the truck. My brother took the 12-gauge pump shotgun while I took the 22-caliber rifle. He would drive us out to one of his biggest fields with the most cattle on it and we started shooting prairie dogs. We were having a blast and must have been shooting for a few hours when my brother started getting a headache. We stopped the truck to attend to him when we noticed that his nose was bleeding and he had two black eyes. It seems that whenever he shot the gun, he would rest his head on the stock and fire. Well, every time he fired that gun, the recoil would hit him right between the eyes. After a very long laugh between the four of us, Dad's buddy asked us to pick out a cow. My brother and I just looked at each other, wondering how shooting prairie dogs all of a sudden switched to cows. He said that we were going to have the freshest steaks we have ever had tonight for dinner and the cow we picked out was going to provide those steaks. So, we picked out a nice fat one and with one shot, Daddy brought it down. We hauled it to the butcher barn, where Dad and he butchered the cow. My brother and I knew that Dad was a butcher before going into the Army, but when we came to find out, Dad's buddy was too. He cut us up some nice two-inch-thick cuts of meat, and we put them on the grill once we got home. Mom and Dad's buddy's wife cooked us a big spread to include real mashed potatoes with brown gravy, fresh corn on the cob from their garden, with fresh green beans also from their garden. And they made yeast rolls and some of the best banana pudding you have ever eaten. Now, this was a feast that could never have been fresher or tasted so good. Oh yes, we stuffed ourselves, but it was good.

While in Ft. Sam Houston, we had a few pets. One that we really remember was a big rabbit, about 20 pounds of rabbit, that we called Herman. Herman was a trained rabbit that would scurry all over the house. He was so trained that we put papers around the toilet, and he would do his business with the papers. One day, we were all watching television when we heard a very loud scream. It sounded like a woman screaming for her life. Tracked the screaming to the laundry room. It seemed that Herman had gotten behind the washer and bit a hole in the hot water pressure side of the washer, and the extremely hot water was just squirting all over him, burning the fire out of him. We finally got him out of there and taped the hose up but after that, all Dad wanted to do was to eat Heman!! Another pet we had was a baby possum. This little guy fell out of a tree in the back yard and my little brother found him. This guy was so tiny that we had to feed him with a syringe. Well, my brother got very attached to him and would take him to school under his hat. One day, he got out and was running all over the school. Well, $300 later, the pest control guys found him. They charged the school $300 to fine him. OH yes, little brother got his ass whipped and so did I just because. Dad said that

I had to be involved somewhere in this and it was better to whip both of us.

My brother and I both became Boy Scouts too. I did not progress as much as my brother. My brother became an Eagle Shout and I used my artistic talents to help him get that far. I can remember making him an Indian headdress out of blue feathers and lots of beads. He had completed his 21 merit badges, with my help and was awarded his medal and badge at one of our Jamborees, where he did his dance wearing the headdress we made when we got his award.

We would go on camping trips and to Boy Scouts to many Jamborees. Again, being from the South, we loved to be outdoors sleeping in a tent and cooking over a fire. I can remember one Jamboree where we were involved in a competition. There were about 10 to 12 different troops involved, and one of the tasks was to build a fire and burn it through a rope that was three feet off the ground. The first to burn through the rope was awarded points toward the Best Troop Award for the Jamboree. We had our team lined up with the other teams. As we started to cut our wood and pile it up for the rope burn, we sneaked in some wood that we had soaked in lighter fluid. At the time, Dad was a heavy smoker, and he had a lighter that you would pull apart and squirt the fluid into the bottom. So, we took the fluid and coated some of our wood. We piled our wood up and lit it and we set a record for burning through the rope. The judges did not suspect our dirty deeds, and we never told anyone about them. Also, during this Jamboree, we brought some items out of our MREs (Meals Ready to Eat) that Dad gave us. We would take some jellies, which were in small cans, and some cheese sauce also in small cans and walk by some scouts that were sitting by their campfires and toss a couple of cans in their fire and walk away. Within five minutes, you would hear some scouts screaming and running around. The cans would explode, scatting the jelly or cheese all over everyone sitting around the campfire! Burned them pretty good too!

Also, at Ft. Sam Houston we found a water plant. The water plant was a big room about 40ft by 40ft square where fresh, clear water was pumped into it and cumulated in a round motion in the room. Well, my brother and I came across the room, and it was hot as hell in Texas. We decided to go swimming in the big room. Now the water was moving pretty fast, and as you got into the water, it would throw you around the room. You had to keep swimming to keep from being sucked into the bottom pipe. You had to swim as fast as you can to keep from being sucked under. There were rails on the wall that you could grab onto to keep from drowning. Yes, another life-or-death situation.

In Texas, it floods a lot. One year my brother and his buddy decided to jump off a bridge in a blue three-ring kiddy pool. The water was flowing pretty fast, and they both got into the pool and jumped off the bridge. They hit the water pretty hard but managed to stay in the boat or pool. They were flying along through the water and came to a bunch of trees with overhanging limbs. As they were traveling under the limbs, a three-foot water moccasin fell out of the tree and landed on my brother's back and bit him in the neck. He was rushed to the hospital, where the doctor said he was a very lucky person because the snake bit him with a dry bite. A dry bite is when a poisonous snake has no poison in the bite. He was sick for a few days and once he was better, Dad whipped his ass OH and my ass too.

At the base, there was and artesian well that supplied water for Salato Creek Park, which was located in the middle of the training part of Ft Sam Houston. It had a pipe about five feet long sticking out of the ground at about a 45-degree angle. An artesian well was under pressure, and the water was just pumping out very forcefully. The water was very cold and very clear. My brother and I would swim in the water, jumping out of the overhanging trees. We also did some fishing around the pipe. We fished it a lot but could not catch anything. We could see the fish in the clear water, but none would bite. So, we took our pellet guns and started shooting the fish. We did pretty well till the SP (Security Police) saw us and arrested us, again. Of course, Dad had to go before the post commander again and explain why we were discharging firearms on his base without permission. I'm not sure what all happened with the new commander but Dad got off with a warning. Once he got home, he whipped our ass but good!

While living in base housing, you had to keep your house and the surrounding yard nice and pretty. They had a program that once a quart, you could win an award. If you win, you will get this big old sign put in front of your house. My brother and I were put in charge of cutting the grass and pruning the flower beds. Mom loved to work in the dirt so we had all these beautiful plants growing everywhere. We had to pull weeds and old plants and it really looked good. We used a push mower to cut the grass; we did not have a gas motor. We were pushing along pretty well, taking turns cutting and pulling weeds. After a few hours, we realized it was easier to pull the mower than push it. Well, Dad saw us do that and he came running out of the house like a bat out of hell. He was screaming at the top of his lungs that that was not the way to cut the grass and if he ever saw us pull that mower again, he would whip our ass. Well, you guessed it, after he went back inside, we pulled it a few times and he came out with his belt in hand and whipped our ass!

My brother and I were both deep into sports during our time at Ft. Sam Houston. We played football, baseball, ran track, and we even played Team Handball. The best memory I have is playing football. I was always a big kid and I could take a punch. I was a quick learner and I love to just run over folks. I played full back on the JV team and can remember many times how I would run over players and sometimes I even knocked out some.

One game, I was running the ball and there had to be five or six defensive players draped all over me, but I kept going. My little brother just loved it when I did that. He said that he could not see anything but my legs because of all the folks on top of me but the pile kept moving. In one game, I took a screen pass over the middle on a two-point conversion. I made it across the line but got hit low from the right and high from the left. My little brother said that all the guys got up but me. I was lying on the ground pulling out grass because my left leg hurt like hell. I was carried to the bench, where they put ice on it. After the game, my knee was twice its size. I was taken to the hospital and Army hospital where they said I messed my knee up pretty good. I was put in the hospital ward. Now, this was not a normal hospital ward. This was the time when Vietnam was going strong. Ft. Sam Houston's hospital was also one of the best in the nation at treating amputees. So, I was put in the amputee ward with all these young men who had arms and legs cut off. I will never forget the cries and screams that echoed throughout the ward. Remember, there were no rooms, so everyone was divided by a cloth divider. You had no privacy. So, all the screams would echo through the room. There were about 40 or so beds in the ward and all the beds were full. The man next to me had lost his left leg below the knee. He would drink a lot of water and I just could not figure out why he drank so much water. One day I ran out of water and he was with the doctors so I leaned over and drank down about half a glass of water. Well, it was not water; it was vodka. I guess due to the pain of losing a leg that he had vodka to help him along. My leg was put into traction with about 5lb weight to keep it straight. Remember this was in the late 60s and medicine was not as it is today. After a few weeks, a nurse went around giving sponge baths to the patients. She was a very beautiful looking young nurse and she came to me and asked if I wanted a sponge bath. Well of course, I wanted one and she told me to roll over on my stomach to where she started to rub hot water on my back with warm cloths and sponges. Now, I am a young 16-year-old with a very nice-looking nurse rubbing my back so my hormones kicked in to the point I had an erection. She asked me to roll over on my back so she could rub my front. I rolled over and just smiled as my erection was on full display. She just smiled, bent over and flipped the head of my penises with her finger. I screamed like a little girl and my erection went away immediately. I figured she must have developed this technique and used it numerous times on all the GIs that were on the ward. After my recovery, my little brother thought it was cool to give me a nickname. It seems while I was recovering, he had told all of our friends that the doctors had to shave my stomach and legs as well as my testicles. So, he gave me the nickname "bald balls." Once I was fully recovered, I beat the holey shit out of him! To this day, this is one of his favorite memories, not that I broke my leg but that I was named "bald balls"!

My brother and I played a lot of sports, normally playing all year round. During track season I would throw the shot put and discus and run the mile; to keep in shape for football. I was the bigger brother, about two feet taller than my brother and he would do the pole valve. At this time, he was in the 8th grade. He never thought anything about the pole valve but due to his size and ability it was the only thing he thought he could do. He would watch all the other folks doing the pole valve and finally figure it out. In Texas, you will have district meets and if you place in the top six or throw good, you can qualify for the state meet. About the second time my brother cleared the bar, he set a district record of 10 feet 6 inches. This height qualified him for state, where he finally finished the meet at a height of 11 foot 2 inches. This was the first time at Cole

High School that an 8th grader lettered in track with a district record. Oh man, I am so very proud of him.

While playing sports in Texas, where it is very hot and humid, you were always in a jock strap. And you would sweat something fears. You would always get a little moister around your testicles. It gets so hot that you develop a rash of sorts, which is called crotch rot. Well, my brother and I found a way to where you can get some relief from this very painful development. We found that if you powered your testicles before practice and after you took a shower, it felt better. But we also found out that it develops crotch biscuits. A crotch biscuit is where the powder around your testicles will ball up into small, flat egg-shaped cakes, which resemble biscuits. These biscuits didn't hurt you, but they sure made practice a little bit harder because they would irritate your junk!

While we played football in Texas, we came up with a way to get back at the starters. The starters thought that their "shit doesn't stink" and that they could get away with whatever they wanted. This pissed off a lot of us and we wanted to get back at them. So, before practice, one day, we broke into the lockers of the starter and smeared what is called Cramergesic

Quick Rub, otherwise known as Atomic Bomb, all in their jock straps. It was an ointment that was used to heat up your body parts that were having cramps. We weren't sure exactly whose strap we put it in until practices started. But once we got going, we knew because the boys were just running and screaming. The coaches didn't know what was going on. It appears that the Atomic Bomb is named that way because it really heats up to help remove the soreness in your muscles. They all ran to the water hose, thinking that the water would make it stop burning but the balm was made to repel water, which made it worse. We had to run laps for the rest of practice till we spewed our guts out, but it was fun watching the starter boys burn!

Chapter 8
Panama

During my senior year and my brother's junior year, my Dad was given an assignment to the Canal Zone, Panama. That had to be the worst time of my life because I had been at Ft Sam Houston, Cole High School for three years and was looking forward to graduating with my fellow classmates. But off to the Canal Zone, we went. It turned out that Panama was the best place my brother and I could have been for that time in our lives. We did so much during those three years that it really changed our life and how we looked forward.

Our first bad encounter in the Canal Zone was when Dad came to pick us up at the Airport. He had to get quarters, Base Housing to live in before we, his depends, could travel. He got base housing within 4 months and had come to get us at the Airport. At that time, we had a wenny dog named Suzy. She had been a part of our family for over 5 years and was a little overweight but was so well-loved. Well, Dad came to the gate to pick us up and we all walked back to the car. Dianna, my little sister, opened the back door to the car and Suzy fell out, dead. Dad had left her in the car with the windows down but the temperature was well into 90's and she died from heat exhaustion.

While we were in Panama, our grandfather on my mom's side passed away. Mom wanted to go to the funeral, so Dad signed up to take a hop on a military aircraft. A hop is a short way of saying Space-Available (Space-A) travel. There are aircraft flying around the world for a number of things, and most will have a few seats available for passengers.

Now, this is not your normal seating arrangement either—it can range from a regular seat aboard a commercial flight contracted by the military to a flight that has just jump seats. Military personnel and their families can fly very cheaply using Space-A. This trip cost Dad $30. It uses a priority system that categorizes travelers by urgency and eligibility. Since Dad was flying on emergency leave because of the passing of a family member, he was the highest priority for travel.

So, before they all left for Arkansas, Dad told us we were *not* to have any parties while they were gone—and if we did, and he found out about it, there would be total hell to pay! Well, that's basically like saying, "Have a party, but don't get caught."

Once my brother and I knew they had left Panama, we started inviting folks over. Everyone had to bring their own drinks. We ended up with about 20 people and numerous ice chests full of beer and liquor. Every time someone finished a beer, we'd put the can on the dining room table. We moved the table against the wall and built a 10-foot-high pyramid out of all the empty beer cans. It was so cool, we took pictures of it—of course, with Mom's camera.

Well, they returned home about 10 days later and never knew a thing about the party—until Mom developed the film. OMG, once Dad saw the pyramid in the photos, he was *furious*! He came into our bedroom, held up the pictures, and asked if we knew anything about them. Of course, we said nothing... and then he took off his belt and started wailing on our butts. My brother and I had so many bruises we couldn't sit down for a week.

We did a lot of fishing in Panama—from using our little yellow raft to fishing from the bank, to fishing on the U-Boat from the Sea Scouts. The best fishing we did was on the U-Boat. We would travel out for miles looking for tide lines. When the tide rises, it pushes water against the bay mouth and brings a bunch of baitfish with it. But once the tide turns and falls, the baitfish get stuck in the tidewater—and that's where the action happens. Large fish feed on the trapped baitfish.

So, we'd drive our boat along the tide line and catch tons of fish. It wasn't unheard of for us to catch 30 to 40 large Spanish mackerel—or even a swordfish or two. We'd have all four deep-sea poles going at once. We also rigged up a scallop trap that we'd hook to the crane on the boat. The trap was a metal cage about five feet long with an opening on the side. We'd let it sink to the bottom, drag it for a half hour or so, and then crank it up. We'd catch four to five-gallon buckets of scallops in a day.

But once you catch scallops, you have to clean them. I remember Dad telling us that since he and the captain did all the work catching them, *we* had to do the cleaning. So, we'd sit on the bow of the boat with our knives and buckets of scallops, cutting them open, taking out the meat, and dropping it into a bucket of ice water—while Dad and the captain drank beer.

We also did a lot of snorkeling. One trip, my brother and I, along with our best buddies, took the train to the Cristóbal side of the canal. The water on that side was always clearer than the Balboa side. We were off to get ourselves a mess of lobsters. We brought two ice chests—one for the lobsters and one for the beer.

Once we got to the beach, we found a really nice coral reef about 8 to 10 feet deep—perfect for lobster hunting. My brother and his buddy took off first and started snorkeling the reef. My buddy and I sat on the beach drinking beers, watching their snorkels bob in and out of the water.

After a while, we saw the two of them zigzagging through the water, and eventually they came running out of the ocean with four really nice lobsters. They told us a 4-foot barracuda had been chasing them. Of course, my buddy and I didn't believe a word of it.

Now it was our turn. We swam out to the reef and started looking down the 8-foot coral walls for two long antennas sticking out. Once we saw them, we'd dive down and try to pull the lobsters out before they locked in with their legs or shot out of the holes.

We were doing pretty well when we noticed a large fish circling us. You guessed it—it was the barracuda my brother saw. My buddy, being Catholic, wore a gold chain with a St. Christopher medal. When you're swimming, it dangles and catches the light, making it look like a lure. And yep, it was attracting the barracuda.

Barracudas are vicious predators. They're not afraid of anything. The fish started getting *really* close, so we decided it was time to head back to the beach. After that, we packed up our lobsters and headed home.

While in Panama, my brother and I fished as much as we could. One time, we were out on the Amador Causeway on the pilot captain's pier. This was where local captains would take a boat out to ships waiting to pass through the canal. They used local captains because they knew how to navigate the canal and Gatun Lake. The pier was over 100 feet long and strong enough to support cars or trucks.

My brother and I loved night fishing, so we went out to the end of the pier and dropped our lines in the water. One night, an old Volkswagen bus drove out onto the pier, pulled up along the rail, and opened the side door. An old man stepped out, took out four fishing poles, and cast them into the water. He clipped them to the rail near the van's door and placed little copper bells on the rod tips.

Then he dropped a big light over the side, stopping just above the water, and lowered a square net below it. After setting everything up, he climbed back into the van and went to sleep.

My brother and I thought this was *the coolest thing in the world*. We kept fishing and kind of forgot about the old man—until we heard the bells ringing. He woke up, set the lines, and reeled in a big fish. He turned on the light above the net, pulled the fish into it, lifted it up, tossed the fish in a cooler, and went right back to sleep. He did that all night long.

My dad was assigned as the First Sergeant while we were in Panama. He was the highest-ranking non-commissioned officer still working directly with the troops. He was responsible for making sure the soldiers met Army standards, managing pay issues, overseeing professional development, and handing out disciplinary actions. It's said that a First Sergeant is the father of the unit—he knows everything and works nonstop.

One day, my brother and I were with Dad at work, and he asked if we wanted to see one of his troops feed his pet boa constrictor. Apparently, this guy had a four-foot boa in a glass cage in the building's basement, and it was feeding day.

Of course, we wanted to see it.

When we got downstairs, the troop was holding a white mouse by the tail and dangling it above the cage. Within seconds, the boa struck and ate the mouse—but it also latched onto the troop's hand. The snake engulfed the mouse *and* part of the guy's hand—his right thumb ended up in the boa's mouth.

It took two guys to hold the snake down and another guy with a pry bar to open the boa's jaws. Once the hand was free, we got to look at it. Around the thumb were a bunch of deep teeth marks, and the guy was bleeding pretty badly. He had to go to the hospital for stitches and shots.

But man, it was so cool to see that snake latched onto a soldier!

While in Panama, Dad would normally go out with the guys on Friday nights. Of course, he would not return till well after midnight and totally drunk. He also would give my brother and me grief about our fishing, about how we never brought any fish home, and he didn't even think we were fishing at all. Well one Friday, my brother and I went fishing. That night, we caught a Stingray about four feet from fin to fin. Once we had the ray on the pier, we developed a plan. We decided to take the fish home and seeing that it was getting close to midnight and Dad would be returning home soon we decided to play a little joke on our dear old dad. We had a carport where the door entered the house in the kitchen, and this was the way our father got back into the house at a late hour. We took the ray and propped it up at the door and put a stick in its mouth to where it was wide open and about knee high. We waited till we saw Dad's car come around the corner of our street to where we hid from view but could see the ray at the door. Well, Dad pulled into the carport and got out of the car. He then headed to the door at the kitchen entrance and before he could get his keys out of his pocket, he saw this monster at the door with those big teeth just shining. Now Dad was drunk as all get out and this is where we heard a little shriek and he just took off running to the front of the house to the main door, running up the sidewalk. But around the sidewalk to the front door, there were these hedges about three feet high to where he hit the first one and went head over ass into the front yard and just slid along the grass. As he was getting up, my brother and I went to the carport door, gathered up the ray, stashed it in the back yard and went to bed. The next morning, Dad came into our room and asked if we knew anything about the monster that he saw at the back door. We played it off, saying we did not know anything about it but he got the story out of us to where he whipped our ass. This one was well worth the ass-wiping for the lifelong memory of Dad flying over the hedges into the front yard.

With the climate the way it was in Panama, six months of the dry season and six months of the rainy season, you learn to adapt. At our base at Fort Clayton, there was this enormous drainage system or a water conveyance system that would flood every rainy season. My brother and I came up with this brilliant idea of sliding down the culverts and seeing where it would take us. We jumped into the raging water that had to be 8 to 10 inches of water with our trash lids and off we went. Took us through the jungle, through and under numerous black palm trees. We, the boys, and Black Palms had a few run-ins during our time in Panama. These trees are armed with slender, brittle spines that can penetrate flesh and go through your flip-flops too! Learned the hard way on this one. They break off in your foot or whatever part of your body it comes in contact with and fester, forming big red pus bags that ooze a white discharge. Why do we know this? Many times, we have had altercations with black palms, and it always seems to leave its mark on your body. As we were riding the waters under these palm trees, we made damn sure that we did not run into them, bump into them, or brush against them. We were traveling along at a pretty good clip when it opened out into a bigger canal, still moving pretty fast so we just kept going. We must have traveled a mile or so, it seemed, till we could see part of the French Cut. The French Cut is part of the old Panama Canal that was started by the French, but due to diseases and death, they stopped cutting into the countryside, leaving a canal that goes nowhere. As we approached the cut, the canal took a 10-foot drop down to the water. We must have flown a few hundred feet into the canal from the speed that we were traveling. From this, we came up with another idea: why not see who can slide on your feet down the counter to the water and jump into the canal? Let's just say that I was not the one to do this. I bet I busted my butt so many times that it knocked some sense into me to the point where I just watched my brother do it. He would ride that green mossy covert like a surfer and jump out into the canal. Fun to watch but there was no way I was going to bust my butt again. We came to find out that during the dry season, when the water went down, we found a barbed wire stretching across the canal. Luckily for us, we only did the slide during the rainy season; otherwise, we might have been gutted like fish.

With our discovery of the French Cut, we found a new playground to explore and investigate. The French Cut was started by France back in the 1800s. They were trying to contact the Pacific and Atlantic oceans so that cargo ships would shorten their travels around the coast. But due to deaths and funding, the French abandoned the Cut and the United States took it over. At school my brother and I had heard about "The Cut," and everyone was saying that it was the best party place to be. Of course, we had to go and check it out. The "place to be" was at one of the highest points in the Cut. The cliffs stood at about 200 feet, and the thing to do was to jump off of the cliff and into the canal below. Well, if you have been reading this book and paid attention, you know that "big brother," yours truly does not mix well with jumping off of things. I did it one time and face planted myself in the water. Oh, everyone there got a good laugh out of it except me. I thought that I had ripped my face off and torn open my stomach; that is how much it hurt. After that one time, all I did was drink beers and talk to the young ladies that were there; all the while, my little brother just kept showing off. Doing flips and swan dives and just showing off.

While at the French Cut, we got to talking to some of our classmates and discovered that there was another local hang-out that was in the interior of Panama not too far from the Canal Zone itself. This was what everyone called Goofy Falls. The Canal Zone is a strip that runs from the Atlantic side to the Pacific side. It is run by the Pan Canal Folks, and outside the wire, as we call it, is Panama. If you were just out driving around the countryside, out where the cows were in the green pastures, you would never know it existed. Goody Falls was a deep cavern cut out by the rainwater, creating smooth rocks and beautiful little pools that went on for a few miles. You had to jump the farmer's fence and walk a few hundred yards and there it was. You could not see it from the road so you had to know where to go to find it. We would take our girlfriends and most of the football team out to Goofy Falls and slide down the smooth black rocks and land in the cool, clear pools of water, drinking beer all the time. We would spend hours just riding the water and drinking and only leave when the beer ran out. This had to be one of the most unknown, beautiful, and pleasing palaces in Panama. As you cross the field, you would come across the still steaming cow pates and you would notice in some of the dried ones, that there were these beautiful looking mushrooms growing out of them. Now, that adventure is for another book! Also, we drank a lot of beer! It was $5 a case and I had my mom's NCO club card. And being a very large person, they never carded me! The folks we partied with were very well off and got most of their beer from their family.

It seems no matter where we were, we did a lot of fishing. They called the Canal Zone Panama the poor man's Hawaii. Surrounded by salt water and fresh water, *fishing* is the best in the world. We discovered that in order for the ships to pass through the canal, a local pilot had to navigate the canal and its numerous locks, including the giant lake called Gatun Lake. At the north and south sides of the canal are a series of locks with Gatun Lake in the middle. A local pilot captain would be boated out to the ship and exchange places with the captain on the boat. There is a fishing pier on the Panama City side of the canal that houses the pilot captains, and you can fish from it. The pier has to be two or three blocks long and goes out pretty deep into the ocean. The causeway out to the pier is called the Amador Causeway, which leads out of Fort Amador. My brother and I would live out on this pier. There are so many stories of fishing out on the causeway. One of those stories is when we were out fishing on the pier and wanted to get our bait out further into the ocean. We started talking to some of the pilot captains and we made a deal with one of them. We would give them a six-pack of American beer, and he would take our line out with him when he went to take over the ship and wait to go through the canal. He took our line out about

300 yards and dropped it in the water. It wasn't a half hour later that we got a bit, and sure enough, we were hooked into some kind of monster. We fought for over an hour, breaking one of our big deep-sea rods into but we got the fish to the bank. We didn't have a big enough net to land him, so we had to walk down the pier and down to the rocks, and we finally got him in. It was a five-foot hammerhead, and this was a female who gave birth to five baby hammerheads right there on the rocks. Oh man, that was one experience that we will never forget.

On numerous occasions, we would fish for what we called Moon Fish. A fish that has the color of the moon and if you put a light on them, they glow. They are about as big as your hand and put off a ferrous fight too. Also, we would catch saltwater gear. They are long, skinny-looking fish that are about three to four feet long and have a mouth full of teeth. We would get out on the rocks on the causeway, bait our hooks with bacon and throw it out into the water. Once the gar gets it in his mouth, you have to wait a few minutes; if you don't, he will throw the hook. You set the hook after a while and let the games begin. These boys will jump out of the water and tail walk for yards at a time.

We didn't eat them, but catching them was such a blast. It is like catching little marlins! Most of the time, we would trade the fish for buckets of fresh shrimp from the local fishermen.

While in Panama, my brother and I became Sea Scouts. The Sea Scouts had an old Army J boat, and we can remember many trips out at sea on it. There were times we would go fishing for scallops. We had made a net that would ride along the bottom and collect the scallops. Once we caught them, we would sit on the top deck and clear them. Nothing better than eating fresh scallops. We did a lot of fishing off of the old J boat too. We would set out four deep sea rods and two hand lines that run alongside the boat, the lines being about 10 feet long. The hand lines were made out of a parachute cord that had a bungee cord in the middle to give the line some flexibility. On one trip while fishing for Spanish Mackerel, we caught a six-foot Wahoo. We caught about 20 mackerel when one of the hand lines went crazy. I was the first to get to the line and started pulling on it. My brother saw that I was having a hell of a time getting it in and he joined in with me. After about an hour of soaking wet, we got the big boy on board. That wahoo was one of the prettiest fish we have ever caught! And it tasted fantastic!

We did some exploring while out on the J boat, too, and we are always looking for better fishing grounds. On one trip, we were boating around and came across a Panamanian shrimping boat. We watched them for a few hours as they brought in their shrimp nets. Most times, the nets were full of shrimp too. So, Dad decided to go over and have a talk with the Panamanians. They were a nice crew and we all got talking about fishing stories. Dad and us boys were drinking beers and telling stories when one of the Panamanians asked if we could make a trade. Well, my dad, being the barter that he is, was all about that. We ended up trading a six-pack for a 5-gallon bucket of fresh shrimp. We ended up with four buckets of shrimp for a case of beer. Beer in the Canal Zone would run you about five bucks a case and that is for the good stuff, so we made out like bandits.

Not only did we do deep sea fishing and fishing from the pier, but Gatun Lake also has some of the best fishing for Peacock Base. Gatun Lake is a large freshwater artificial lake that forms a major part of the Panama Canal. There are locks at both ends of the lake, which were formed by building the Gatun Dam. It is a part of the canal, allowing ships to pass through the locks and into the channels in the lake. Local pilots will board the ships and navigate the channels to the other side. Within the lake, there is prime restate or habit for the peacock bass to thrive. My brother and I took full advantage of this opportunity to catch these fish. We are not sure how we acquired a yellow heavy-duty blow-up dingy, but we had one and fished in it all the time. We think that Dad did some more bartering, acquired it, and gave it to us. There were so many trips out to parts of the lake where we could launch the boat and paddle out into the lily pads. But before you could go fishing, you had to get some bait. We developed a method on how to get our bait which our bait of choice was minnows from the local steams that surround our house. We would take a wine bottle that has a concave bottom. Knock a hole in the bottom and place some bread in it, sink it in the stream and put the cork back on. Within a few minutes, you could have anywhere from just a few minutes to a full bottle. The best time to catch fish, we found out, was when it rained, and to fish the lily pads. It was not uncommon for us to catch seven or eight peacock bass, weighing as much as five or six pounds. We would catch fish till it stopped raining or we ran out of bait. Panama has a six-month dry season and a six-month wet season so you have plenty of time to catch fish! We found that the best way to eat our fish was to bar-b-que them on the grill or to make ceviche. Ceviche is normally served as an appetizer, but we often eat it as a main meal. It's made with raw, fresh fish marinated in lemon and lime juice—usually for about 15 minutes—then spiced with chili peppers, red onions, salt, cilantro, and other seasonings to taste.

With both a lemon tree and a lime tree in our backyard, and plenty of fresh peacock bass, we eat a ton of ceviche.

Grilling the fish is simple: take a fresh, cleaned whole fish, smear it with your favorite sauce, wrap it in foil, and throw it on the grill. Once the fish is flaky, it's time to chow down!

Another adventure was when we were out drinking. We were out in some pretty open water catching fish when we noticed the channel markers, which let the boat captains know where the deep water is so that they can navigate through the lake, and we decided to grab ahold of one. We had a few beers and decided to have a little fun, not even thinking about how we were going to get back to the shore; see that once we grabbed the channel marker, our boat got away from us. We climbed up on the marker, which was about five feet tall, and started "mooning" all the boats that came by. We were butt naked out in the middle of the lake, not a care in the world, and "mooning" everyone that came by. We got a lot of "cat whistles" and a lot of other words yelled at us but we didn't care at all. After a while, a big boat came into the site and started heading toward us. We pulled our shorts up and just waited for them to reach us. It turns out that the big boat belonged to the Governor of Panama and they decided that they were going to help us get our boat back once we got to the shore. The Governor was pretty cool about the whole thing, telling us that we should not be mooning all of Panama; all the time, he was smiling and trying to hide his laughter. We got back to shore, got our boat, thanked him for his help, and proceeded to drink more beer.

As time went on in Panama, we made some pretty good friends. Some of these friends owned little islands out in Gatun Lake. One trip my brother and I really remember was when the whole football team went out to one of these islands and camped out overnight. We took cases of beer and hamburger buns and meat and our fishing gear to include innertubes. Once we got out to the island, there was this little hut on the island where we could store our gear, and it had a grill out to the side. This is where we found out that beer floats. We would get in the innertube, put a few beers between your legs and start catching fish with the beers floating between your legs. We also found out that if you hooked a big enough bass, it will tow you across the lake. I hooked into a monster and sure enough, I was off on a trip. The good thing was that the floating beers stayed with me. I was floating along, drinking my beers all the time being towed by a monster fish. Remember I had mentioned what we brought to the island? Well, we did not bring any water, nor did we bring any lighter fluid, but we did have beer! We ended up eating our hamburger raw or, as some folks called it, steak tartar.

Another wonderful thing that happened during the sleepover on the island was something that my brother and I will never forget. We had all heard of this island, which was nearby and owned by the Governor of Panama. One of the biggest rope swings I had ever heard of was on this island. So, we all piled into our 12-

foot John Boat with a 25-horsepower Johnson motor and took off. There were five of us in the boat with a very large Igloo cooler of beer. I was driving the boat and my brother was at the front, navigating us around the other islands. We had such a load in the boat that there must have been only two inches of area between floating and sinking so I was going at a very slow pace of course, drinking beers as we went. At this time of the year, there were numerous boats out on the lake. As boats passed us, they created some pretty good wakes, some as high as three feet, and we took on water every time, bailing as we went. All of us were scoping water and throwing it out just to keep us afloat. A few times, the water created by the other boats was pretty high and I just slowed down to elephant speed so as not to capsize the boat. Lucky for us, we did not sink but luck was not on our side when we came into some four-foot wakes. There seemed to be numerous boats pulling skiers that day and of course, as a skier, you always want to get close to things, including other boats and spray them with your skis. This one jerk got so close that he soared over all of us, but that is when the fun or not-so-much fun started. I was going as slowly as I could, and after every wave, we kept taking on water- lots of water. About the third wave, I told everyone that we were not going to make it through the next wave. We went up to the crest of the wave and came straight down, swamping the boat. The funny part about this is all five of us were yelling, "Save the beer" as we went down. The boat flipped over completely to where we were all hanging on to the bottom and all of us were holding onto the Igloo cooler. We were all busting out laughing, and it finally hit us. It was just how we were going to make it to the island. One big boat saw us in distress and came over and asked us if we needed a tow. It turned out that that boat was going to help us was the governor's boat, the Governor of Panama and that they were on their way to the Governor's Island. They hooked up a rope and towed us to the island. We turned the boat over, got all the water out of it and started looking for the rope swing, drinking beer all the time. Once we found it, it had to be all three to four stories high with a three-inch braided rope with two-by-fours at the end. You would swing out at least 50 yards and let go. When it came to my turn, I had a little problem. I got a good run and took off, getting to the end of the swing where I should have let go, but for some reason, I did not let go. So here I am, coming back into the tree at 100 miles per hour and all the guys are telling me to let go. I didn't let go till I hit the tree, falling about 20 feet to the ground, where I was met by these giant root formations. These root systems had the dirt between them worn away, worn down by the water to where the root was sticking out about two feet and looked like the veins in your arm, branching out everywhere. I bet I hit five or six of them before settling down near the water's edge. The guys were laughing so hard and telling me I needed to do that again. Luckily for me, I didn't break anything but I had some pretty big burses from the fall.

Many folks reading this book might want to know how the name Austini came about. Our last name is Austin so who or what put an "i" in our name? Well, at the end of the football year, there is a game between the winning school against and all-star team made up of all the other teams. And there is also a coach's selection of all the best players in their position. It seems that my brother and I, even though we knew we were good, were selected as the best at what we do. So, the newspaper does a write-up about those players and when it came to Austin, the writer just wrote, "And the outstanding play by the Austini brothers" enabled the

Canal Zone Jr College to go undefeated. From then on, we were known as the Austini brothers. Loved it!

My brother and I were sports freaks, and the best time we had was when we played sports together. You cannot put into words playing side by side with your little brother. The feeling we both shared during those times will never be forgotten! One of those experiences was when we played Junior College football together. I was an offensive tackle, and my little brother was a tight end. On defense, we played inside defense tackles. So, when we lined up on the line of scrimmage, we were always side by side. During our last year together in college, I was selected as one of the team captains so I was responsible for keeping the guys in line.

Yes, we were a little cocky but we were really good. We went undefended that year. There were some wonderful memories of the Austini brothers playing football. One was when we had just been given a warning from the officials to watch our langue. We had just scored a touchdown and we were lining up for the extra point. It was so quiet on the field. As we approached the line of scrimmage, my left guard just happened to say a few words to the other team about how bad they were of course, using explicate language. The whole stadium heard what he said and busted out laughing and the official threw the flag. I got the team together and told them that if we kept using bad language, we would keep getting a penalty and eventually, we would be kicked out of the game. So of course, as we approached the line again, the whole team told the other team how they felt using explicate language. This again cost us five yards, and I got an earful from the officials. On the third time, we came to the line of scrimmage, we all kept our mouths shut and kicked the extra point.

All of us on the team had played together throughout high school, so we were close. One game, the coach and all the folks on the bench were giving us grief because we were whipping their ass. So, when we huddled up for the next play, I called a play. Being the team captain, I called sweep right toward the other team's bench, but I said that we were going for the first down but to keep running and run over the coach and all the players on the bench. So, the whole team, all eleven of us, including our quarterback, pulled down the line of scrimmage and just plowed over the coach and bench players. Of course, my coach called me over and ripped me a new ass but the team got a kick out of it. Of course, we drove the ball down the field and scored and won the game. The team never forgot about that game, and I gained so much respect for it.

On many occasions my brother and I would line up in positions where we both would line up on both sides of the center on defense. This kind of formation was called a split-six formation. You would have six down linemen and two linebackers. We had very strong and fast linemen and two of the best linebackers in the league. My brother and I developed a method that would scare the other team. As the quarterback got under the center, my brother and I would time it just right and jump offsides and plow over the center and quarter back and just pound them into the ground. After a few times, we would fake it and the center would jump and fall down. Every time we did this, it would cost the other team five yards. We loved to do this throughout the whole game.

There were times when the football team had to travel to the other side of the canal and play the other side. We were on the Panama City side, southside and we had to travel to the Colon side on the north side of thc canal. We would take the train that travel from side to side of the canal. There was a time when my brother and I found out that a player on the other team was being considered for a scholarship to a big school back in the States. We knew who he was and we didn't think that he was any better than we were. We decided that we were going to show him how the "Austini" brothers played. He played defensive end and I played left tackle. My brother played tight end. Every few plays, he would line up between me and my brother in the gap. So, we put our plan together. I was going to hit him low, below the knee, and my brother was going to hit him high, above the waist. Once the ball was snapped, we went after him. When the play was over all we could hear was a shriek and we saw him rolling around in pain on the turf. It seems we kind of messed up his knee and he never returned to the game. Of course, now in football, this kind of play is not allowed but back in our time, it was.

On one occasion, we were traveling by train. These trains were the mass transportation that carried folks from one side of the canal to the other side. One trip we took was very memorable to me and my brother. As the captain of the football team, I had some power over the players. As we were going through the mountains and woods of the interior of the Canal Zone, I saw the other train coming at us on the outside tracks.

I instructed the team to hang their butts out the window when the oncoming train approached; we were going to moon the whole train. So, there was a box car filled with about 30 or so football players with every window filled with butts hanging out. We just laughed afterward. After the little incident, I was called in front of the coaches and got my ass ripped and instructed on how to correctly deal with the team. As I walked away, I could hear the coaches all cracking up!

Some other experiences that did not involve sports while we were stationed in Panama were the times we had a 1969 Volkswagen Karmann Ghia. It was blue with a black convertible top. Now, this was not your average Karmann Ghia; no this was a VW that had been in Panama for years and it showed it in many areas of the vehicle. One area was the parking brake or, sometimes called the emergency brake which was located between the driver's seat and the passenger seat. This was a level-type braking system. When you were parked, you would pull up on the handle until the brakes were applied. Well, our brake leveler was on the middle hump between the seats so your pedals were below the brake. The reason for all this info is that our driver's side floor panel that contains the seat rails that let you make adjustments was rusted out. Oh, I mean, it rusted out so much that the rails were the only thing there where you could see the road. Well, in Panama, it rains 6 months out of the year. So, as you drive along and hit a big, deep puddle of water, the driver's side floor will flood. It will fill up with about five to six inches of water. Also, the rail that the seat slides on was just hanging there with no support so my brother and I took a wire hanger and wired it to the emergency brake handle. So, you are driving along during the rainy season and hit a puddle. You hold your feet up out of the water till you pass through the puddle, when all the water drains out and then you put your feet back on the gas pedal and the clutch pedal and just drive on.

Another thing about our 1969 Karmann Ghia was that the starter didn't work all the time. My brother and I didn't have time to work on the car because we had other important things to do, mainly drinking beer with our buddies. So, when we would go out on a date, sometimes together, we would try and park the car on a hill. When the starter didn't work, we would tell our girlfriends to get out and push because we weren't going to push; someone had to drive and be the co-pilot. But one of the most fun stater experiences was when we parked the Karmann Ghia in front of the house and had to go to school early in the morning. The funny part was looking at two high school students sitting in the car and their mom in her curlers and house coat pushing the car to get it started! Thank God it was easy to push and always started on the first try!

We had some good times with that car. There were many times when we had a few buddies with us and needed to go on a beer run. Most of the time, we would go to the Class Six package store located in the NCO Club on the base where we were stationed. The Class Six is a military phrase for a liquor store, and the NCO Clubs would always have a Class Six store. At the time, my brother and I were both underage to buy liquor but to get around that problem, my mom sent me on a mission to get Dad some beer and she let me use her membership card. Will it seem I never gave it back here? I kind of forgot about it on purpose. Being a pretty good-sized teenager, the clerk never carded me. So, in our little Karmann Ghia, we would pack in 5 football players and a large cooler of beer and go off to The Pub made of two box cars. The Pub was two old railroad cars on old tracks. The old rails have been long removed except for the tracks the old box cars are on. They put in two big bars that sell all kinds of liquor and beer on tap, put two 20-foot concrete slaps out front and you have the Box Cars. You would drive your vehicle up to the concrete with your lights facing the center and start partying. When it got dark, you would put your headlights on and keep on parting. Most times, there would be a stage where a local band would play.

While in Panama, my brother and I joined a local rugby team that was composed of high school kids. We would play against each other, and when a ship came through the canal, we would play their rugby team. We played teams from England, Scotland, Italy, and Germany. Now, the teams on the ships were semi or pro rugby players. Normally a ship would take weeks to get through the canal from one side to the other, so there was plenty of time to play a game. We would start with giving a gift to each team, normally a banner or pin. After the exchange, we would beat the holy shit out of each other for a few hours. After the games, we all became best friends. We would go to our favorite place to drink, The Box Cars.

When we played the English team, they showed us how to make and English drink using beer and lemonade from England. This was called a "Lager Top". Come to find out, after playing a few hours in the hot sun and drinking a "Lagar Top," it was very refreshing to the point after drinking two or three, you were to take on the world or drink more beers. We would always have a blast with the different teams, and those memories of the rugby team have stuck with my brother and me for all these years.

Other than beer being so cheap, so was the liquor. Our drink was rum and coke. You could get a half gallon of rum for $3 and the good stuff was only $4. Now, we didn't have your normal size drink; oh no, you drank your rum and coke in a mason jar, quart size. And your rum and coke would consist of filling the jar full of rum and just splashing coke in there for looks. I got so drunk on rum and cokes that, to this day, I can't drink one without getting sick.

Another really cool experience we got to do was the Ocean-to-Ocean race through the Panama Canal. This is a seven-man team that races for three days and starts at Cristobal at the Atlantic entrance of the canal to the Diabol ramp on the Pacific side. It is a 35-mile race. When we did the race, we had canoes made from hollowed-out logs and the paddles were mostly native-style. Most canoes were only about two to three inches above the water, and we all had to have bail cans. The race normally begins on a Friday and concludes late on Sunday. The first leg starts on the Atlantic side, ending at Gatun Lake. This is a short leg that can have some of the biggest waves and many sandbanks to negotiate but it is mostly a straight race. The second leg is the most difficult because you have to travel through Gatun Lake. It begins at the Catun locks and ends at Gamboa. This leg is 20 miles. This can be the most beautiful part of the race because you travel the length of Gatun Lake. Gatun Lake is a major part of the Panama Canal. This is where ships will travel the length of the lake to go through sets of locks and both ends. The lake has ships that go through the Panama Canal, and you can paddle right next to these large vessels. The final day of the race consists of three stages. The first stage is a short sprint race, the second is an hour race through the Culebra Cut which is about 8 ½ miles with the final stage of another hour ends at the Miraflores locks to Diablo port. We could see the Miraflores lock from our house at Ft. Clayton. The most memorable time of the race is when you get into the lock systems. In total, there are 12 locks you have to pass through. You had to tie up your canoe to the wall. The water in the locks can drop from only five or six feet to as much as 28 feet. Many times, you would be in the lock with a ship transferring the Panama Canal. The best memory of the race was the party before the race started. We would set up our canon and fill it with beer. The canoe was about 15 feet long. We would have all kinds of local foods and drinks and party all day, which made the next day of racing a little bit harder.

My brother and I have moved on in life. It has been five decades now. We have gotten married twice or more, and both have had very successful careers. My career was in the military for 20 years, followed by 22 years working for the Department of Defense in Logistics. I have traveled the world in the military and have been involved in many of our nation's disputes, from the rescued of American hostages in Iran, where I was in charge of the ground recovery group, to rescuing hostages in Grenada, where I was responsible for the vehicle recovery, to vehicle ground support for the involvement of Gadhafi, to being deployed in

Operation Desert Shield and Desert Storm. My military career led me to all parts of the world.

My brother's career was 32 years in the Department of Defense in Corrosion, painting military aircraft on a Naval installation in Florida. After enlisting in the military for four years, he started his career in the Department of Defense. You might have seen some of his work if you had watched the Blue Angles perform! He has painted numerous aircraft, from fighter aircraft to bombers to helicopters. He has a photo of each and every one he has painted and has them on the wall at his house. He developed into the number one painter of aircraft and became a trainer for those who came through his shop.

We both had two children, a boy and a girl, and my brother is a grandfather as well. We call each other at least twice a week and talk about our lives and the lives we live. We ask each other if we would change anything in our past and we agree that we would not. Even though our bodies are broken and scared from our numerous adventures, we would not change a thing because it made us the men we are today. As we have gotten older, our bodies have been reminding us of so much pain. Our adventures have built character, personality, and temperament and molded us. It created us into who we are, and we would not give that up for anything.

We both call our mom every weekend, I call on Saturdays and my brother calls on Sundays and Mom really looks forward to talking to her boys every week. We both wish we could spend more time with Mom, but we both live so far away that it gets hard to make the long trip.

We talk to our little sister at least once a week to catch up and to get the real story about how our Mom is doing. Our little sister has become the caregiver for Mom since Dad passed. She lives only 30 minutes from Mom. How day our sister and mom are like too crazy a young woman, taking adventures every chance they can. Mom is pushing 92 years old now and likes to keep a positive swing on her life and doesn't like to give us the "whole story." By calling our sister and asking her about Mom, we get the whole picture. We like to say, "Tell a phone, tell a graph, tell a Dianna," because she can't stop talking. We will do this until we die; this will never change. This is our family, and we believe that family is the most important thing. You need to keep a strong family tie.

Gallery

...lions and pass interceptions setting up scoring opportunities for the opponents. Also their inability to sustain a successful offensive drive to score inside the opponent's 20 yard line. Cristobal High School must maintain their scrappy opportunist playing to have a shot at interscholastic championshiip. Regaridless of the statistics, Canal Zone College maintain an odds on favorite due to the experience and overall size of the squad. The outstanding line play of the Austini brothers, linemen May, Wolf, Dekle and Robinson gives the College the line blocking necessary to gain yardage and score the necessary points for victory.

Two brothers, one reputation.
The kind earned through mud, sweat, and silence.

Tiger-Dogs Will Not Be Easy Pickings For College Devils

The Canal Zone College in a repeat of the 1973 season again won the 1974 interscholastic football championship. Improving over 1973 performance the "Green Devils" posted a perfect record of 4 wins, no losses. In posting a 4-0 record, the Canal Zone College allowed a total of 14 points through the entire season scoring a total of 57 points.

The greater percentage of credit of this outstanding record must be given to the splendid offensive and defensive line play of the "Green Devils". This is quite evident in the selection of 4 interior linesmen to the All Zone offensive team and 5 selected to the All Zone defensive team. Many of these boys saw both offensive and defensive action playing most of the games without relief.

The "Green Devils" may find the going in the Palm Bowl quite a bit tougher than during the regular season. The ability of the all stars to field separate offensive and defensive teams could have a physical bearing on the game with fresh troops in the game on each tournover of the ball.

There is a good possibility that the "Green Devils" will open up in the Palm Bowl breaking away from their conservative power type football with the use of a passing attack along with their powerful ground game. The all stars with more ___ and ability to keep

team, there will be an excellent half-time program. The Balboa Elementary School with a group from each grade level, 1 through 6, will demonstrate the building of pyramids. Mesdames Dawn and Wendy Stairette, Captain and Co-Captains respectively of the Cristobal High School Majorettes, will give an outstanding performance each using two flaming batons.

The Palm Bowl Committee again wishes to thank all the men, women, and students for their unselfish giving of their

time in selling tickets, ushering, the donation of shirts and trophies by Abernathy, S. A. and Palm Bowl tickets by Dalton Travel Agency in making this memorable United Way Palm Bowl.

Introduction of players, coaches and the officials who are donating their time to the Palm Bowl game will get underway at 6:45 p.m. Immediately following the introduction, the captains of the two squads will flip the coin for ball possession and the kickoff scheduled for 7 p.m.

Palm Bowl Teams

CANAL ZONE COLLEGE "GREEN DEVILS"

No.	Name	Pos.	Weight	Grade
20	McAfee, David	HB	145	Freshman
22	Brown, Richard	HB	144	Sophomore
30	*Gibson, Jay	HB	155	Sophomore
32	*Young, James	FB	175	Sophomore
33	Cobb, Keith	QB	140	Sophomore
43	Gordon, Jaime	HB	155	Freshman
67	Bowman, Ronnie	G	175	Sophomore
52	Robinson, Frank	C	160	Freshman
60	Wulf, Philip	T	175	Sophomore
58	Gilbert, Robert	G	155	Freshman
64	Doble, John	G	180	Sophomore
65	Koehler, Joe	C	170	Sophomore
66	May, Mike	T	170	Freshman
68	Dahn, Donald	G	175	Freshman
73	*Austin, David	T	210	Sophomore
74	Fish, Jim	T	210	Sophomore
76	Padro, Jorge	T	175	Freshman
80	Dekle, John	E	180	Sophomore
83	Jaramillo, Reinaldo	E	145	Freshman
87	Matthias, Alvin	E	148	Freshman
88	Austin, Don	E	160	Freshman

*Team Captains

Coaches   Cheerleaders   Managers

A newspaper clipping highlighting the Canal Zone College "Green Devils" football team, featuring a full team roster and game preview.

Putting in the work long before game day.
Some lessons came with sweat, not words.

Where toughness wasn't taught — it was tested.
And bonds were built in the dirt.

Shown above are the Balboa High School Bulldogs, 1973 champions in Interscholastic Track and Field competition.

(Left to right): Front: Gordon, Quezada, Small, Brown, Green, Ray, Arana, Tobinson, MacAfee, Rankin, Morrison, Goldman.

2nd. Row: Asst. Coach Husten, Rivera, Mennard, Connell, Barthlett, D. Austin, Shields, D. Austin, Cobb, Mgr. Cook, Coach Cleve D. Oliver.

Back Row: Trey, Lowe, Wohlman.

Champions on paper, but the real win was in every lap,
lift, and lesson.
1973. Balboa. We ran with heart.

Names on a list, numbers on jerseys — but the grit was personal.
Zone champs. Brotherhood in pads.

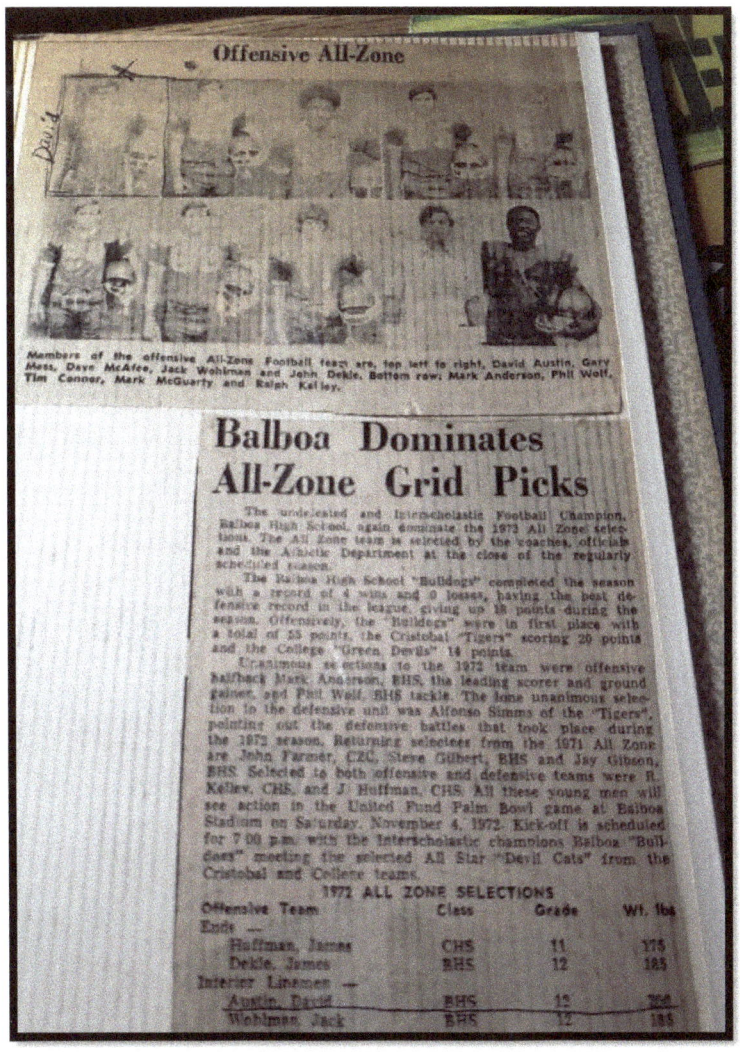

Offensive All-Zone

Members of the offensive All-Zone Football team are, top left to right, David Austin, Gary Moss, Dave McAfee, Jack Wohlman and John Dekle. Bottom row, Mark Anderson, Phil Wolf, Tim Conner, Mark McGuarty and Ralph Kelley.

Balboa Dominates All-Zone Grid Picks

The undefeated and Interscholastic Football Champion, Balboa High School, again dominate the 1972 All Zone selections. The All Zone team is selected by the coaches, officials and the Athletic Department at the close of the regularly scheduled season.

The Balboa High School "Bulldogs" completed the season with a record of 4 wins and 0 losses, having the best defensive record in the league, giving up 18 points during the season. Offensively, the "Bulldogs" were in first place with a total of 55 points, the Cristobal "Tigers" scoring 20 points and the College "Green Devils" 14 points.

Unanimous selections to the 1972 team were offensive halfback Mark Anderson, BHS, the leading scorer and ground gainer, and Phil Wolf, BHS tackle. The lone unanimous selection in the defensive unit was Alfonso Simms of the "Tigers", pointing out the defensive battles that took place during the 1972 season. Returning selectees from the 1971 All Zone are John Farmer, CZC, Steve Gilbert, BHS and Jay Gibson, BHS. Selected to both offensive and defensive teams were R. Kelley, CHS, and J. Huffman, CHS. All these young men will see action in the United Fund Palm Bowl game at Balboa Stadium on Saturday, November 4, 1972. Kick-off is scheduled for 7:00 p.m. with the Interscholastic champions Balboa "Bulldogs" meeting the selected All Star "Devil Cats" from the Cristobal and College teams.

1972 ALL ZONE SELECTIONS

Offensive Team	Class	Grade	Wt. lbs
Ends —			
Huffman, James	CHS	11	175
Dekle, James	BHS	12	185
Interior Linemen —			
Austin, David	BHS	12	200
Wohlman, Jack	BHS	12	185

Recognition came in ink — but the respect was earned in silence.
All-Zone. All grit. All heart.

Family

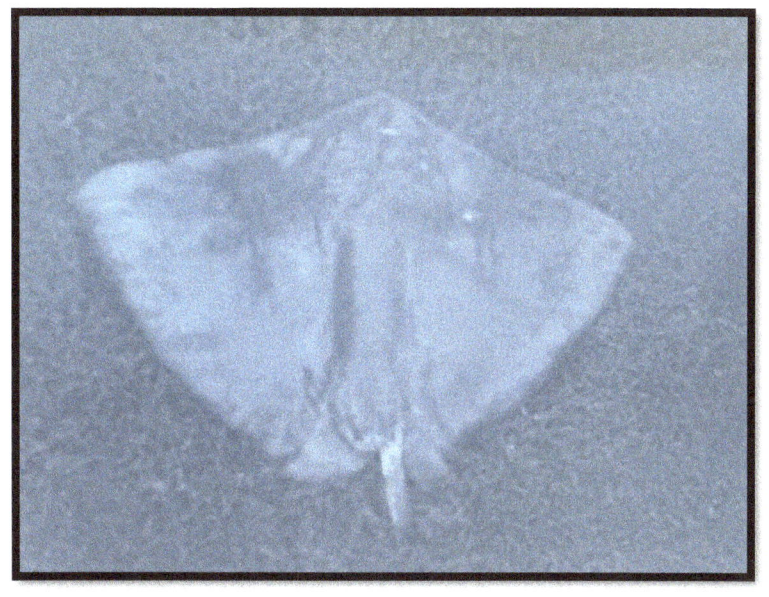

About The Author

I'm very proud to be a military brat! My father served in the Army for 27 years, and both my brother and I lived the military life for over 18 years. We are both United States Air Force veterans, and now our children are military brats too.

Being a military brat shaped who we are. I spent 20 years in the USAF and traveled the world, but nothing compares to the days I spent growing up with my little brother—those were some of the most rewarding days of my life.

I wrote this book for two main reasons: first, to share a good story, and second, because I know there are millions of military brats out there who also have stories to tell. I spent over eight years writing this book, much of it reminiscing with my brother about our shared experiences. We'd spend hours on the phone talking about the "good old days"—from fishing trips and football games to which country we were stationed in. It's amazing how just a few words can bring back so many vivid memories.

I truly hope this book resonates with readers and helps spark memories of their own unique journeys.